The Nature and Use of
Ecotoxicological Evidence

The Nature and Use of Ecotoxicological Evidence

Natural Science, Statistics, Psychology, and Sociology

Michael C. Newman
Virginia Institute of Marine Science, Gloucester Point, VA, United States;
College of William & Mary's Virginia Institute of Marine Science,
Gloucester Point, VA, United States

ACADEMIC PRESS

An imprint of Elsevier

Academic Press is an imprint of Elsevier
125 London Wall, London EC2Y 5AS, United Kingdom
525 B Street, Suite 1800, San Diego, CA 92101-4495, United States
50 Hampshire Street, 5th Floor, Cambridge, MA 02139, United States
The Boulevard, Langford Lane, Kidlington, Oxford OX5 1GB, United Kingdom

Notices
Knowledge and best practice in this field are constantly changing. As new research and experience broaden
our understanding, changes in research methods, professional practices, or medical treatment may become
necessary.

Practitioners and researchers must always rely on their own experience and knowledge in evaluating and
using any information, methods, compounds, or experiments described herein. In using such information or
methods they should be mindful of their own safety and the safety of others, including parties for whom they
have a professional responsibility.

To the fullest extent of the law, neither the Publisher nor the authors, contributors, or editors, assume any
liability for any injury and/or damage to persons or property as a matter of products liability, negligence or
otherwise, or from any use or operation of any methods, products, instructions, or ideas contained in the
material herein.

British Library Cataloguing-in-Publication Data
A catalogue record for this book is available from the British Library

Library of Congress Cataloging-in-Publication Data
A catalog record for this book is available from the Library of Congress

ISBN: 978-0-12-809642-0

For Information on all Academic Press publications
visit our website at https://www.elsevier.com/books-and-journals

Working together
to grow libraries in
developing countries

www.elsevier.com • www.bookaid.org

Publisher: Candice Janco
Acquisition Editor: Erin Hill-Parks
Editorial Project Manager: Tracy Tufaga
Production Project Manager: Vijayaraj Purushothaman
Cover Designer: Christian Bilbow
Cover Photograph: Michael C. Newman

Typeset by MPS Limited, Chennai, India

Dedication

To my wife, Peg, and sons, Ben and Ian.

... all is but a woven web of guesses.

Xenophanes (530 BC).

Contents

Section 3
How Groups Weigh and Apply Evidence

Section 4
Conclusion

About the Author

Michael C. Newman is currently the A. Marshall Acuff, Jr. Professor of Marine Science at the College of William & Mary School of Marine Science, where he also served as Dean of Graduate Studies from 1999 to 2002. Previously, he was a faculty member at the University of Georgia Savannah River Ecology Laboratory. His research interests include quantitative ecotoxicology, environmental statistics, risk assessment, population effects of contaminants, metal chemistry, bioaccumulation and biomagnification modeling, and during the last 15 years, qualities of innovative concepts and technologies that foster or inhibit their adoption by the ecotoxicology scientific community. In addition to more than 150 articles, he authored five books and edited another five on these topics. Mandarin and Turkish translations of his *Fundamentals of Ecotoxicology* are available from Chemical Industry Press (Beijing) and PALME (Ankara). His marine risk assessment book was translated into Mandarin in 2011. He taught full semester and short courses at universities throughout the world including the University of California, University of South Carolina, University of Georgia, College of William & Mary, Jagiellonian University (Poland), University of Antwerp (Belgium), University of Joensuu (Finland), University of Technology Sydney (Australia), University of Hong Kong, University of Koblenz-Landau (Germany), Huazhong Normal University (P.R. China), and Royal Holloway University of London (UK). He served numerous international, national, and regional organizations including the OECD, US EPA Science Advisory Board, US EPA ECOFRAM, US EPA STAA, and the US National Academy of Science NRC. He was a Fulbright Senior Scholar (University of Koblenz-Landau, Germany, 2009) and a Government of Kerala Scholar in Residence/Erudite Scholar (Cochin University of Science and Technology, Cochin University, Kerala, India, 2011). In 2004, the Society of Environmental Toxicology and Chemistry (SETAC) awarded him its Founder's Award, "the highest SETAC award, given to a person with an outstanding career who has made a clearly identifiable contribution in the environmental sciences." In 2014, he was also named a SETAC Fellow, for "long-term and significant scientific and science policy contributions."

Preface

INTENT

... I saw everything they have there, and examined it carefully. ... What birds and beasts haven't I seen there! ... What flies, butterflies, cockroaches, little bits of beetles. And what tiny cochineal insects! Why really some of them were smaller than a pin's head!

But did you see the elephant? What did you think it looked like? I'll be bound you felt as if you were looking at a mountain.

... Well, brother you mustn't be too hard on me; but, to tell the truth, I didn't remark the elephant.

Krylov (1814/1869)

In the above exchange from Ivan A. Krylov's fable, *The Inquisitive Man*, a museum goer relates the marvelous minutiae that he viewed while unintentionally missing the largest item on exhibit. This book attempts to redirect environmental scientists' attention to such an elephant looming over the innumerable bits of engrossing knowledge gathered together in the science of ecotoxicology.

After more than 35 years researching and communicating pollutant effects, I am convinced that the most serious impediments to wise action are the misconstruing of evidence by the scientific community and miscommunicating evidence to regulators and the public. How ecotoxicologists amass, vet, and share evidence is swayed by the cognitive psychologies of individual scientists and the social dynamics of scientists. What evidence comes to dominate the exchange among scientists, regulators, and decision makers depends on both scientific soundness and social circumstances. Unfortunately, relevant psychological and sociological factors receive short shrift during deliberations by professionals so unfamiliar with the social sciences that they consider them inaccessible to formal study. Instead of trying to understand and keep in check these psychological and sociological influences, environmental professionals tacitly treat them as impenetrably complex, yet somehow sufficiently constrained.

My experience is that the impact of these influences varies widely among the activities of environmental scientists. Major impediments to the advancement of our understanding of pollutants and to sound decision-making occur often enough to warrant closer scrutiny of the psychological and social circumstances influencing evidence generation and communication. This short book examines this elephant in the ecotoxicology gallery of facts and concepts.

SCOPE AND TENOR

Let not men think there is no truth but in the sciences that they study, or the books that they read. To prejudge other men's notions before we have looked into them is not to show their darkness, but to put out our own eyes.

Locke (1706/1999)

This book necessarily draws on concepts and methods from diverse disciplines, so chances are that most readers at some point will be unfamiliar with the materials being laid out. This unfamiliarity might create a distracting uneasiness for many who routinely focus their professional readings on the natural sciences. To avoid conceptual dissonance, most chapters include brief overviews of the relevant concepts and then focus down on how they are germane to environmental science issues.

Ecotoxicology has been defined in various ways, many of which emphasize the suborganismal (e.g., elevated cytochrome P450 monooxygenase titer) to classic ecosystem (e.g., diminished nutrient retention or biodiversity) levels. In many definitions, humans are considered only as the source of contaminants. Here, ecotoxicology is defined more broadly as the science of contaminants in the biosphere and their effects on constituents of the biosphere, including humans. This is consistent with Truhaut's original (1977) and more recent (Newman, 2015) definitions. The scale of ecotoxicology is also extended to landscapes and ultimately to the biosphere. Relevant contaminants include ecotoxicants such as excess nitrogen in coastal waters and excess greenhouse gases in the earth's atmosphere that disrupt ecological systems but are not usually included in discussions of environmental toxicants. Radioactive contaminants are included where pertinent.

It is important to also state what will not be covered. The intriguing topic of deliberate scientific fraud is excluded for two reasons. The first reason is that no ecotoxicology examples come readily to mind except the well-worn science bending embedded in the current climate change debate (McGarity & Wagner, 2008). The second is that, in my opinion, the cumulative impact of many unnoticed foibles is much more harmful to the science than are the occasional fraudulent acts.

Specific examples are essential for highlighting foibles and suggesting ways of minimizing their impacts. Every effort was made during discussions of perceived shortcomings to obscure the identity of any associated individuals. Unfortunately, complete obscurity was impossible to achieve for a few committees in which particularly illustrative behaviors seem to have emerged. When concealment was not successful or possible, the individual or group being discussed is reminded that this book claims to present informed opinion alongside generally accepted fact. Every effort was made in wording to differentiate informed opinion from generally accepted facts. To further identify instances of opinion, a personally discomforting style of writing in the first person singular was taken for the first time in this author's career. As with scientific concepts generally, any interpretation—such as an opinion in this book—holds favored status as a plausible explanation only until further evidence requires its displacement or modification. In contrast to attempts to obscure the identities of those displaying foibles, every effort will be made to identify participants involved in exemplary generation or communication of evidence. The restrained approach to foibles is motivated by civility and that for excellence by frank admiration.

COVERAGE

This book has four sections: Introduction, How Individuals Gather and Judge Evidence, How Groups Weigh and Apply Evidence, and a brief Conclusion. The Introduction sketches out the history of pollution and the reasons why timely and sound evidence is now absolutely essential for human well-being. The large section about individuals starts with a chapter on general human reasoning, especially common mistakes. It is followed by two smaller chapters that examine how individual scientists reason, and perhaps, make errors in the process. The fourth chapter in this section focuses on statistical methods that individual scientists use to make inferences. The next large section, How Groups Weigh and Apply Evidence, broadens coverage from microlevel interactions to those at a macrolevel or group interactions, particularly as they influence evidence-based judgments. Topics in this section include network qualities of copublishing scientists, movement of novel concepts and techniques into groups of scientists and expert panels, and finally, decision-making by large Internet-defined collections of individuals. In Chapter 7, "How Innovations Enter and Move Within Groups" and Chapter 8, "Evidence in Social Networks," respectively, of this last section, presentation will shift to exploring concepts by analyzing a set of ecotoxicology concepts and techniques. The conclusion consists of a single chapter that brings together the most salient points and potential remedies for problems discussed in the preceding chapters.

REFERENCES

Krylov, I. A. (1869). The inquisitive man. In W. R. S. Ralston (Ed.), *Krilov and his fables* (pp. 43−44). London: Strahan and Company Publishers. (Original work published 1814).

Locke, J. (1999). Of the conduct of the understanding. In R. W. Grant, & N. Tarcov (Eds.), *Some thoughts concerning education and of the conduct of the understanding*. Cambridge: Hackett Publishing Co. (Original work published 1706).

McGarity, T. D., & Wagner, W. E. (2008). *Bending science*. Cambridge, MA: Harvard University Press.

Newman, M. C. (2015). *Fundamentals of ecotoxicology. The science of pollution* (fourth ed.). Boca Raton, FL: CRC/Taylor and Francis.

Truhaut, R. (1977). Ecotoxicology: Objectives, principles and perspectives. *Ecotoxicology and Environmental Safety*, *1*, 151−173.

Acknowledgments

The support by Mr. and Mrs. A. Marshall Acuff, Jr., of my endowed professorship is greatly appreciated. Their continued support makes possible projects like this book. I am very grateful to Margaret Mulvey who carefully read and corrected the manuscript prior to submission. Marcos Krull was my technical partner in the ecotoxicology innovation diffusion and co-authorship network studies presented in this book. The insights from those studies would have been far fewer without his exceptional R skills during the last 3 years.

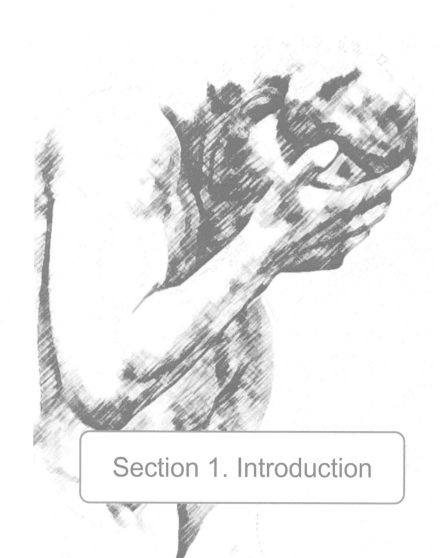

Section 1. Introduction

Chapter 1

The Emerging Importance of Pollution

Before we can draw conclusions about the origins of environmental problems, we need historical accounts of the concrete interactions between society and nature that have produced them. From those histories we can infer the modes of thought and behaviour that are more likely than others to be detrimental to the environment we want to live in. A primary element of such histories should be the social analysis of scientific knowledge construction...

Bird (1987)

1.1 INTRODUCTION

The public's attention was first drawn to the unacceptable consequences of pollution by two extraordinary books, *Silent Spring* (Carson, 1962) and *Minamata* (Smith & Smith, 1975). Underlying each book is the disquieting truth that enough evidence existed in the scientific community to have avoided the catastrophes altogether. Relative to proper pesticide use, Carson (1962) points out that, "Much of the necessary knowledge is now available, but we do not use it." A century before methylmercury ruined the lives of dozens of Minamata children, its extreme toxicity was widely reported in the scientific literature (Iriguchi, 2012). These tragedies did not result from a lack of knowledge: they were a consequence of inattention by the scientific community and inadequate communication of evidence to decision makers and the general public. Given that enough knowledge existed beforehand to avoid the disasters, two central questions emerge: "How could the conditions leading to these disasters have come into existence?" and "Why was existing evidence not gathered together and communicated to decision makers in time to avert the disasters?" Answers, albeit discomforting ones, will emerge in the historical perspective integrated into this chapter.

The Nature and Use of Ecotoxicological Evidence. DOI: https://doi.org/10.1016/B978-0-12-809642-0.00001-7

1.2   HISTORICAL EMERGENCE OF POLLUTION

1.2.1   Pre-20th Century: The Slow Emergence of Pollution

The sulphates [in air] rise very high in large towns, because the amount of sulphur in the coal used as well as of decomposition. . .. When the sulfuric acid increases more rapidly than the ammonia the rain becomes acid.

Smith (1872)

By 1888, sulfuric acid rain from [the Japanese Asio copper] smelter had killed 5,000 hectares of forest and contaminated local waters.

McNeill (2000)

Humanity's unrivaled capacity to extract energy and resources from the environment gives rise to pollution. Such pollution became progressively worse through time as extractive technologies improved and the number of humans applying them increased. Pollution came into being historically when contaminant releases exceeded the capacity of the surrounding environment to dilute or neutralize them.

At first, the import of pollution changed quite slowly in lockstep with humankind's slow increases in population densities, aggregation into cities, and expansion into unoccupied areas of the globe. Unacceptable levels of localized pollution did eventually appear as population growth accelerated.

The world population grew from a mere 14 million in 3000 BCE, to more than 200 million by 700 CE, to a remarkable 1.5 billion by 1900 CE (U.S. Census Bureau, 2016). Although most pre-20th century humans lived rural lives, people began to gather in progressively larger cities. The first cities were modest by current standards. The largest in 3000 BCE, for instance, Uruk and Memphis, contained only tens of thousands of inhabitants. Athens had a population of 300,000 by 430 BCE and Rome in the first century CE had 650,000 (Markham, 1994). Within the first 700 years CE, populations in cities such as Constantinople, Rome, and Chang'an had grown to accommodate roughly a million inhabitants each. By 1850, there were 100 cities of 100,000, 37 cities of 200,000, 9 cities of 500,000, and 3 cities of more than 1,000,000 citizens (Chandler, 1987). By the middle of the 19th century, the world's 49 largest cities combined held more people than lived on the entire earth in 3000 BCE. As the 19th century drew to a close, 14% of the 1.5 billion people making up the world population lived in cities (UN, 1999).

Although widespread pollution outside of cities occurred as populations spread and increased in densities, the first blatantly unacceptable pollution was experienced in cities. The Thames River around which London had grown was unvaryingly rancid by 1860 (Markham, 1994). As reflected in the above quotes, the 19th century likewise brought localized acid rain around large cities and industrial regions. Because materials making up cities warm faster than surrounding lands, large cities also created heat islands that retained pollutants. Polluted city air warmed, rose above the city, cooled, sank to areas adjacent to the city, and was drawn back into the city again to replace new masses of air rising from the city center. The resultant convention cell worsened conditions by retaining polluted city air. Natural weather conditions such as thermal inversions exacerbated urban conditions. Seventeenth century London experienced increased numbers of human deaths during episodic "stinking fogs" created by weather conditions that prevented polluted air from escaping the city (Anderson, 2009; Brimblecombe, 1987).

Change in human population number and density was only one half of the problem. Equally astonishing changes were occurring in our resource extraction technologies, including those associated with food production, domestic heating, and eventually, industrial production. The Holocene agricultural development in the Fertile Crescent approximately 11,500 years ago represented substantial advancement in our abilities to draw sustenance from the environment. However, as early as 3000 BCE, intense demands on land by the Mesopotamian agrarian society led to extensive salinization (Grimm et al., 2008). Similarly, intense agriculture across the Middle East, India, and China between 2000 BCE and 1000 CE brought the first of three major human-induced pulses of soil erosion (McNeill, 2000). A second global surge in soil erosion due to agriculture and animal grazing began during European expansion into the Americas and accelerated with expansion also into Siberia, the Maghreb, South Africa, Australia, New Zealand, and the Northern Caucasus (McNeill, 2000).

Nonagricultural resource extraction intensified through this period. First century CE lead mining in southeast England and Spain by the occupying Romans created broad swaths of contaminated land. Shifts from wood and charcoal to coal occurred at different times in different regions of the globe. To support the increasing demand for iron products, coal mining intensified during the Northern half of the Sung dynasty (960−1126 CE) and peaked at levels only matched centuries later during the 18th century industrialization of Europe (Hartwell, 1962). Coal mining in Europe, and especially England, increased in the 13th century until coal eventually emerged as a crucial domestic, and then industrial, fuel in early cities (Steffen, Grinevald, Crutzen, & McNeill, 2011). Restrictions on domestic coal burning and building of taller factory smoke stacks temporarily reduced city air pollution by permitting pollutants to dilute to acceptable levels over longer distances from the source. These measures proved to be only temporary solutions as cities and industries continued to grow.

The intensity of resource extraction surged dramatically with the Industrial Revolution of the 18th and 19th centuries. By the mid-19th century, extensive land pollution appeared in major regions of the developing world (McNeill, 2000). Intensified mining and smelting activities during the 1868 Meiji Restoration brought widespread contamination of Japanese lands and crops. Mining and processing of coal, iron, and nickel caused similar land contamination in other parts of the world as the 19th century drew to a close.

What is the message to be taken from this sketch of the pre-20th century population growth and resource utilization? Before the 20th century, intermeshed increases in population growth and resource extraction efficiency gave root to the pollution issues of today. Deterioration of environmental conditions proceeded slowly, starting several millennia ago and continuing into the last century. During the 20th century, problems would manifest that required a fundamental rethinking about how humans conduct business on the earth and would eventually stimulate the development of new technologies for coping with pollution. Although occasionally portrayed as a whimsical meandering of modern human sensibilities, changes in our collective mindset and efforts to control pollution begun in the 20th century were necessary responses to indisputable practical problems.

1.2.2 20th Century: Running Up the Bill for the Next Generation

Only within the moment of time represented by the present century has one species — man — acquired significant power to alter the nature of his world.

Carson (1962)

One of the newest fads in Washington — and elsewhere — is environmental science. The term has political potency even if its meaning is vague and questionable.

Klopsteg (1966)

"Present the repair bill to the next generation" became the unspoken slogan of those who exploited nature for short-term gains [in the 1940s].

Kline (2011)

The human population and its capacity to extract resources surged again in the 20th century. The world population reached 2.5 billion by mid-century and topped 6 billion by the end of the century. Although growth rates in some regions of the world were slowing or even falling below simple replacement rates, approximately 78 million people—more than five times the entire human population of 3000 BCE—were added to the global population in the single year of 2000. By then, the urban population of the earth had grown to include 29% of all humans (Grimm et al., 2008; Satterwaite, 2002). The number of large cities increased, especially in North America and sub-Saharan Africa. New York and Tokyo reached the 10 million residence threshold, becoming the first megacities. Three dozen cities would eventually become megacities in the 20th century (Satterwaite, 2002)

Changes in resource utilization were equally extraordinary. Excluding areas of rock, ice, and other barren lands, an estimated 48% of the earth's land was dominated by or partially disturbed by human activities by 1990

(AAAS, accessed 02.06.16). The third great pulse of land erosion began in the 1950s and continues to this day (McNeill, 2000).

Chemical technology also changed so rapidly that the term, Age of Chemistry, has been applied to the mid-20th century advances in the chemical industry and our enthusiastic consumption of its products. This industry drew upon different, and still more, resources to produce its life-improving products, resulting in an increased and more diverse waste stream from society.

Actualized by the Age of Chemistry and other scientific advances, the Green Revolution began in the 1940s and intensified into the 1960s. Global food production was substantially enhanced by methodically integrating high-yield crop strains, chemical fertilizers, and chemical biocides. Fossil fuel−powered farming equipment became a key component of this revolution. An irrefutably beneficial advance for humankind, the Green Revolution, also brought unique resource conservation, worker safety, consumer safety, and general pollution issues that would eventually require redress.

Surges in human population and resource extraction created widespread pollution during this century: our collective environmental debt grew so large that it could no longer be ignored. Although earlier problems had been localized around cities of high population density or in areas of intense resource extraction such as mining or agriculture regions, pollution became globally pervasive during the 20th century. Laws, mores, and scientific enterprises slowly evolved to cope with the accumulated environmental debt and daily costs of drawing imprudently upon our increasingly limited resources. Unfortunately, these changes were not fully instituted as the root causes came into existence for the Minamata disaster and pesticide poisoning of wildlife mentioned at the beginning of the chapter. The Age of Chemistry was well underway by the time Chisso Corporation began discharging mercury into Minamata Bay; but environmental laws were still weak and the science of pollution was in its infancy. The organochlorine pesticides used so successfully to control arthropod vectors of human disease were being applied increasingly to agricultural plots as the Green Revolution boomed. No one was ready to accept the fact that what was so beneficial in disease control might be harmful to wildlife when applied to agricultural production.

Exemplary change agents emerged during this century. Relative to workers' health, Alice Hamilton successfully argued at the beginning of the century that workplace safety practices must include consideration of chemical exposure (Hamilton, 1985). In the latter half of the century, Carson (1962) crafted the extraordinary book, *Silent Spring*, which heightened our collective awareness of the negative consequences of advanced agricultural practices. Gene and Aileen Smith (Smith & Smith, 1975) published an equally impactful collection of black and white photographs depicting the ruined lives of children and adults due to the Minamata mercury poisoning. All catalyzed widespread changes relative to how humankind perceived its wastes. Soon thereafter, René Truhaut broadened his career-long concerns about human exposure to contaminants in food to encompass pollutant impacts on nonhuman species and ecological entities (Truhaut, 1970). Working within the International Council of Scientific Union's Ecotoxicology Working Group, Truhaut (1977) was the first to define the new applied science of ecotoxicology.

Ecotoxicology is the branch of Toxicology concerned with the study of toxic effects, caused by natural and synthetic pollutants, to the constituents of ecosystems, animal (including human), vegetable, and microbial, in an integral context.

1.2.3 21st Century: Budget Balancing While Paying Down the Debt

Until the late 20th century and into the 21st century, humans labored under the false pretense that their actions alone controlled the course of nature and that water and land resources were theirs for the taking. Humankind has mistakenly taken for granted that the present biospheric life support system, in which Homo sapiens has evolved and flourished, has and will always be present.

Cairns (2011)

As environmental issues grow in economic significance and as science takes on increasing importance in influencing public opinion and resolving environmental policy debates, suppression of environmental science has become "increasingly common."

Kuehn (2004)

Demographers predict that the global population will grow at a progressively diminishing rate until a maximum of 10 billion is reached in 2200 (UN, 1999). This transition to a stable population size is not due to some mysterious quality of the human race: the cause should be obvious from the above narrative. The human population is adjusting as it approaches or perhaps overshoots maximum resource extraction rates and the capacity of the environment to cope with its byproducts. These demographic adjustments manifest in social, economic, legal, and political transitions. Social shifts in responses to an early 19th century resource shortage—the Irish Potato Famine—included delayed marriages, fewer children per couple, and modified rules of land inheritance. On a broader scale, the demographics of couples in developed and developing countries shift toward fewer children and more investment in each child as personal income increases. China's one-child policy introduced in 1978 was a clear instance of political change that is now only being phased out, as the Chinese economy flourishes and the demand for youthful workers increases.

Global urbanization continues into the 21st century. By 2000, nearly 50% of world's population dwelled in urban areas, that is, areas with extremely high-resource demands, and consequently, opportunities for intensified pollution (Satterwaite, 2002). Waste discharge from urban areas grew to impact global, in addition to local, biogeochemical cycles (Grimm et al., 2008).

To encapsulate what has been described to this point (Fig. 1.1), the patterns of population growth and resource use that would eventually create today's pollution issues were established long before the 20th century. During the 20th century, the increasingly harmful pollution resulting from the intensification of these patterns manifested with clearly unacceptable consequences. This intensification was of such a magnitude that the period following the World War II (WWII) is sometimes referred to as the Great Acceleration (Steffen et al., 2011). The mercury poisoning of Minamata's children and dichlorodiphenyltrichloroethane (DDT) poisoning of birds emerged mid-century as blatantly unacceptable consequences of our surging industry and agricultural practices. Figuratively, the environmental bill came due and a change in behavior was needed to balance our increasingly large withdrawals from the environment, and to make up for the heretofore, insufficient attention paid to assure a sustainable biosphere. A third global theme, conscientious resource conservation and pollution control, was very slowly being integrated into human behaviors in the 20th century.

Given the history described above, it is impossible to dismiss arguments that vital skills for the 21st century humankind must include effectively acquiring, discerning the utility of, and then applying sound knowledge about our byproducts in the environment. Without these skills, the danger of many more and worse disasters than the Minamata and DDT poisonings of the mid-20th century will arise with increasing frequency.

1.3   EMERGENCE OF THE ANTHROPOCENE

Until recently the historians and the students of the humanities, and to a certain extent even biologists, consciously failed to reckon with the natural laws of the biosphere, the only terrestrial envelope where life can exist. Basically man cannot be separated from it; it is only now that this indissolubility begins to appear clearly and in precise terms before us. He is geologically connected with its material and energetic structure.

Vernadsky (1945)

A global context became increasingly more relevant for pollution management as humankind grew and its activities intensified during the last few centuries. Fortunately, scientific investigations of human influences also expanded to a global context, albeit after a protracted period of wavering within the relevant scientific communities.

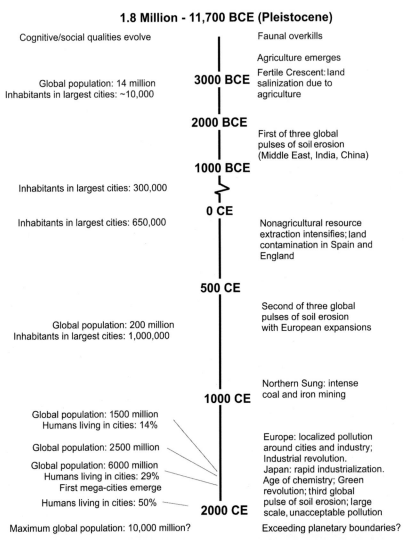

FIGURE 1.1　Timeline summarizing the general trends in human population (left) and human resource extraction achievements and concomitant pollution (right).

The Austrian geologist, Eduard Seuss (1904) introduced, and later the Russian geochemist, Vladimir Vernadsky (1945) elaborated upon, the concept of the biosphere—the integrated biological system that occupies the Earth's troposphere, hydrosphere, and upper layer of the lithosphere. These pioneers saw the biosphere was worthy of study *as a unit*. In his 1945 treatise, Vernadsky reminded the post-WWII scientific community of Aleksei Pavlov's late 1920s assertion that humankind had modified the biosphere enough to warrant naming a new era, the anthropogenic era. Noting humankind's mounting influence in the biosphere, Vernadsky further proposed that a new planet-wide "geological phenomenon" had come into existence, the noosphere. He defined the noosphere as the "reconstruction of the biosphere in the interests of freely thinking humanity as a single totality" (Vernadsky, 1945). Our actions were literally reshaping the biosphere. The international politics of the time, and perhaps a language barrier, delayed Vernadsky's ideas from being more widely disseminated.

The Gaia hypothesis was the next easily distinguished stage in scientific thinking about our planet-wide impact. Introduced in the early 1970s by James Lovelock and championed by Lynn Margulis soon thereafter, the broad Gaia hypothesis is that "life at an early stage of its evolution acquired the capacity to control the global environment to suit its needs and that this capacity has persisted and is still in active use" (Lovelock, 1972). An argument is made that the biosphere regulates global features such as atmospheric composition, surface albedo, temperature, and surface water pH. The Gaia hypothesis came under criticism, perhaps as a consequence of naming a scientific hypothesis after a mythological goddess or the emphasis on life's cybernetic control of conditions within a range favorable to itself. The implication of purposefulness was off-putting for many natural scientists. Narrower definitions were later proposed for the hypothesis (i.e., "organisms and their environment evolve as a single, self-regulating system"; Lovelock, 2003), that permitted formal testing of proposed instances of biotic control. Nonetheless, the Gaia hypothesis established a framework for the study of global processes and pollution issues like greenhouse gases and aerosol impacts on global climate, the impact of pervasive refrigerant release on ozone layer integrity, and more recently, the role of anthropogenic carbon dioxide emissions on ocean acidification.

The current phase of scientific thinking involves a growing movement to name a new human-dominated geological epoch, the Anthropocene (Crutzen, 2002). An 18th century beginning of the Anthropocene is proposed based on elevated carbon dioxide and methane levels measured in air samples entrapped in polar ice. The Early Anthropocene Hypothesis (Steffen, et al., 2011) proposes an even earlier beginning based on the Pleistocene mass extinctions of megafauna on four continents (50,000−10,000 years ago), and later Holocene emergence (115,000 years ago), and further expansion of agriculture into China, the Americas, Africa, and India roughly 7000−4000 years ago (Lewis & Maslin, 2015). Unlike the Gaia hypothesis, the proposal to recognize this new epoch enjoys rapidly increasing support in the scientific community. Zalasiewicz, Williams, Steffen, and Crutzen (2010) provide their explanation:

> ... why did it not also become a discarded footnote in the history of geological ideas? It helps that the term [Anthropocene] is vivid, as much for the public as for scientists. More importantly, it was coined at a time of dawning realization that human activity was indeed changing the Earth on a scale comparable with some of the major events of the ancient past.

Evidence for a distinctive human-dominated epoch include an order-of-magnitude increase in widespread erosion and sedimentation, increases in key atmospheric gases such as carbon dioxide and methane, rises in global temperatures and sea levels, unprecedented rates of species extinctions, and increasing ocean acidity.

Global change has accelerated since 1945 to the point that considerable scientific effort is now being spent to ascertain the planetary boundaries for human endeavors relative to major processes such as climate change, biodiversity loss, influences on the nitrogen and phosphorus cycles, stratospheric ozone depletion, ocean acidification, freshwater use, land use, atmospheric aerosol loading, and chemical pollution (Rockström et al., 2009; Steffen et al., 2011). These planetary boundaries are the safe operating levels for human activities in the biosphere. Human's pursuit of social and economic progress will continue unimpeded as long as these boundaries are not exceeded. Unfortunately, humanity is currently believed to be working beyond the planetary boundaries for climate change, loss of biodiversity, and the nitrogen cycle. Ellis (2011) stresses that our activities relative to these boundaries are expected to accelerate at an increasing pace:

> Industrial human systems are far more strongly connected globally and tend to evolve more rapidly than prior social systems, accelerating the pace of social change, material exchange and tool development, and the tempo of human interactions with the biosphere – a change in the rate of biospheric change that may be novel in itself.

Given the theme of this book, it is disconcerting that the planetary boundaries for chemical pollution are listed as "to be determined" or "not yet defined" in the latest scientific publications on the topic (figure 1.1 of Rockström et al. (2009); table 1 of Steffen et al. (2011)). This uncertainty presents ecotoxicologists with an unprecedented challenge during the 21st century. Human populations will reach 9.6 billion in this century—approximately 686 times the number of people alive in 3000 BCE. Resource extraction will accelerate and novel waste streams will appear such as those from genetically modified organism (GMO) crops, personal care products, veterinary and human medicines, and electronic equipment. As stated above by Ellis (2011), the global reach of changes is broadening and the tempo of change is accelerating. The way in which we generate, evaluate, and then use ecotoxicological evidence needs to evolve in step with this new reality to facilitate evidence-driven solutions to current and emerging perturbations of the natural world.

This book is intended to contribute to this necessary improvement in how we conduct our business as environmental scientists and risk assessors. The issues discussed will be familiar to most readers but the vantage will not be typical for natural scientists. The emphasis will be on the impediments introduced by our cognitive psychology and social dynamics, that is, the elephant in the room discussed in the Preface.

1.4 CONCLUSION

The increase in the size and complexity of human societies has probably not been accompanied by significant changes to our social instincts. While natural selection can sometimes lead to substantial genetic change in a few thousand years, most biologists think that important changes in complex characteristics take much longer to assemble. Our innate social psychology is probably that bequeathed to us by our Pleistocene ancestors.

Richerson and Boyd (2005)

Our inherent capacity for eking out a living evolved during the Pleistocene epoch and has remained essentially unchanged since then. Modern social interactions are shaped by behaviors evolved to maximize cooperation in small, genetically related groups and to minimize intergroup conflict with other, genetically distant groups competing to varying degrees for the same environmental resources. Relative to inherited reasoning skills, Richardson (2007) speculates that, as long-lived and large animals with wide home ranges, early humans had to adapt cognitively to a broad range of circumstances varying over large geographical distances and long time periods. Rapid climate change during the Pleistocene also required enhanced cognitive abilities in early humans. Others speculate that sociality and attendant language acquisition were also major forces in shaping the human capacity to reason.

Dawkins (1982, 1987, 1989) argues persuasively that a second major replicon in addition to the gene, the meme, emerged during the Pleistocene. A meme is "a unit of cultural inheritance, hypothesized as analogous to the particulate gene, and as naturally selected by virtue of its "phenotypic" consequences on its own survival and replication in the cultural environment" (Dawkins, 1982). Many (e.g., Blackmore, 1999; Brodie, 1996; Lynch, 1996) accept the meme concept and the role of meme complexes (memeplexes) that together create an enhanced fitness. Genes are transmitted vertically, that is, from parent to offspring. With the novel capacity for symbolic representation as exemplified by language, human brains retain memes that can be transferred vertically and also horizontally, that is, among individuals in a group. This second type of replicon has greatly accelerated humankind's abilities to grow and extract resources from its environment. Memes also permit rapid modulation of our genetically determined foibles. Whatever the evolutionary reasons, there is no denying that cognitive flexibility and meme transmission remain remarkably advantageous to modern humans.

Evolved qualities advantageous to small groups of hunter-gatherers living on the Pleistocene plains of Africa cannot be expected to be optimal in all aspects of our 21st century lives. Indeed, academicians who specialize in cultural evolution have identified several Pleistocene adaptations as maladaptations in modern life (Richardson, 2007; Richerson & Boyd, 2005). The attribution of many modern suboptimal cognitive and social behaviors to once-adaptive Pleistocene traits is called the "big mistake hypothesis"[1] (Burnham & Johnson, 2005). Basic heuristics inherited from our distant ancestors allow modern humans to efficiently navigate through their everyday lives; however, these same heuristics are inadequate in situations like estimating the risk of harm from an environmental pollutant. Metaphorically speaking, the same cognitive and social qualities inherited from the Pleistocene that put humans on the moon also caused the Space Shuttle Challenger disaster.

Many modern institutions have developed ways of reinforcing useful social behavior and modes of reasoning, and coping with the more problematic ones. College business courses explore how group dynamics influence productivity, with discussions ranging from information exchange dynamics to the maximum number of individuals that one person should directly supervise. When attempting to resolve long-standing problems in communities, aid agencies consider the predictable dynamics of groups exposed to novel ideas. Rogers (1995) gives the example of a failed program to encourage the boiling of drinking water in a Peruvian village because public health workers failed to engage the correct village opinion leaders. As a last instance, medical practitioners are beginning to learn how to best communicate to their patients the risks and probabilities associated with lifestyle choices and diagnostic test results (Gigerenzer & Gray, 2011).

Extraordinary hubris would be required to argue that inherited human foibles do not manifest in attempts by ecotoxicologists to understand and cope with pollution, and that consideration of these foibles is not an essential best practice in ecotoxicology and environmental risk assessment. Since the Pleistocene, humans with diverse specialties have attempted with varying degrees of success to use accumulated knowledge to moderate the influence of inherited maladaptive traits and shibboleths. The extraordinary challenges of 21st century pollution and population require recognition of maladaptive behavior and meme diffusion be incorporated into the new science of ecotoxicology.

---

1. This hypothesis relates specifically to fitness of cooperative or altruistic traits associated within small Pleistocene human groups made up of closely related individuals. The hypothesis focuses on how altruistic traits can be maladaptive in many modern interactions within groups of unrelated individuals; however, similarly suboptimal fitness in cognition can also be imagined for people attempting to cope with the environmental issues described earlier.

REFERENCES

AAAS: American Association for the Advancement of Science. *Atlas of population and environment*. (2001). <http://atlas.aaas.org/index.php?part=2&sec=landuse&sub=luintro> Retrieved 06.02.16.

Anderson, H. R. (2009). Air pollution and mortality: A history. *Atmospheric Environment, 43*, 142–152.

Bird, E. A. R. (1987). The social construction of nature: Theoretical approaches to the history of environmental problems. *Environmental Review, 11*, 255–264.

Blackmore, S. (1999). *The meme machine*. Oxford: Oxford University Press.

Brimblecombe, P. (1987). *The big smoke: A history of air pollution in London since Medieval times*. London: Methuen & Co., Ltd.

Brodie, R. (1996). *Virus of the mind*. New York: Hay House, Inc.

Burnham, T., & Johnson, D. (2005). The biological and evolutionary logic of human cooperation. *Analyse & Kritik, 27*, 113–135.

Cairns, J., Jr. (2011). A call for humility and an end to humankind's disruption of evolutionary processes. *Integrated Environmental Assessment and Management, 8*, 2–3.

Carson, R. L. (1962). *Silent spring*. New York: Houghton Mifflin Co.

Chandler, T. (1987). *Four thousand years of urban growth*. Queenston: St. David's University Press.

Crutzen, P. J. (2002). Geology of mankind. *Nature, 415*, 23.

Dawkins, R. (1982). *The extended phenotype*. Oxford: Oxford University Press.

Dawkins, R. (1987). *The blind watchmaker*. New York: W.W. Norton & Co.

Dawkins, R. (1989). *The selfish gene* (2nd ed.). Oxford: Oxford University Press.

Ellis, E. C. (2011). Anthropogenic transformations of the terrestrial biosphere. *Philosophical Transactions of the Royal Society A, 369*, 1010–1035.

Gigerenzer, G., & Gray, J. A. M. (2011). *Better doctors, better patients, better decisions*. Cambridge: The MIT Press.

Grimm, N. B., Feath, S. H., Golubiewski, N. E., Redman, C. L., Wu, J., Bai, X., & Briggs, J. M. (2008). Global change and the ecology of cities. *Science, 319*, 756–760.

Hamilton, A. (1985). Forty years in the poisonous trades. *American Journal of Industrial Medicine, 7*, 3–18.

Hartwell, R. (1962). A revolution in the Chinese iron and coal industries during the Northern Sung, 960-1126 A.D. *The Journal of Asian Studies, 21*, 153–162.

Iriguchi, N. (2012). *Minamata Bay, 1932*. Tokyo: Nippon Hyoron Sha.

Kline, B. (2011). *First along the river*. Plymouth, UK: Rowman & Littlefield Publishers, Inc.

Klopsteg, P. E. (1966). Environmental sciences. *Science, 152*, 595.

Kuehn, R. R. (2004). Suppression of environmental science. *American Journal of Law & Medicine, 30*, 333–369.

Lewis, S. L., & Maslin, M. A. (2015). Defining the Anthropocene. *Nature, 519*, 171–180.

Lovelock, J. E. (1972). Gaia as seen through the atmosphere. *Atmospheric Environment, 6*, 579–580.

Lovelock, J. E. (2003). The living Earth. *Nature, 426*, 769–770.

Lynch, A. (1996). Thought contagion. How belief spreads through society. New York: Basic Books.

Markham, A. (1994). *A brief history of pollution*. London: Earthscan Publications, Ltd.

McNeill, J. R. (2000). *Something new under the sun*. New York: W.W. Norton & Co.

Richardson, R. C. (2007). *Evolutionary psychology as maladapted psychology*. Cambridge: MIT Press.

Richerson, P. J., & Boyd, R. (2005). *Not by genes alone. How culture transforms human evolution*. Chicago: University of Chicago Press.

Rockström, J., Steffen, W., Noone, K., Persson, Ä., Chapin, F. S., Lambin, E. F., . . . Foley, J. A. (2009). A safe operating space for humanity. *Nature, 461*, 472—475.

Rogers, E. M. (1995). *Diffusion of innovations* (4th ed.). New York: The Free Press.

Satterwaite, D. (2002). *The scale of urban change worldwide 1950-2000 and its underpinnings. International Institute for Environment and Development. RICS International Paper Series*. London: Royal Institute of Chartered Surveyors.

Seuss, E. (1904). Das Antlitz der Erde *(The face of the earth). (H. B. C. Sollas, Trans.)*. Oxford: Clarendon Press.

Smith, R. A. (1872). *Air and rain. The beginning of a chemical climatology*. London: Longmans, Green & Co.

Smith, W. E., & Smith, A. M. (1975). *Minamata*. New York: Holt, Rinehart and Winston.

Steffen, W., Grinevald, J., Crutzen, P., & McNeill, J. (2011). The Anthropocene: Conceptual and historical perspectives. *Philosophical Transactions of the Royal Society A, 369*, 842—867.

Truhaut, R. (1970). Survey of the hazards of the chemical age. *Pure and Applied Chemistry, 21*, 419—436.

Truhaut, R. (1977). Ecotoxicology: objectives, principles and perspectives. *Ecotoxicology and Environmental Safety, 1*, 151—173.

U.N. (1999). *The world at six billion, ESA/P/WP.154*. New York: United Nations.

U.S. Census Bureau. *Historical estimates of world population*. (2016). <https://www.census.gov/population/international/data/worldpop/table_history.php> Retrieved 24.01.16.

Vernadsky, V. I. (1945). The biosphere and the noösphere. *American Scientist, 33*, 1—12.

Zalasiewicz, J., Williams, M., Steffen, W., & Crutzen, P. (2010). The new world of the Anthropocene. *Environmental Science & Technology, 44*, 2228—2231.

Section 2. How Individuals Gather and Judge Evidence

Chapter 2

Human Reasoning: Everyday Heuristics and Foibles

...a large majority of the general public thinks that they are more intelligent, more fair-minded, less prejudiced, and more skilled behind the wheel of an automobile than the average person. This phenomenon is so reliable and ubiquitous that it has become known as the 'Lake Wobegon effect' after Garrison Keillor's fictional community where 'the women are strong, the men are good-looking, and all the children are above average'.

Gilovich (1991)

It ain't what you don't know that gets you in trouble. It's what you know for sure that just ain't so.

Attributed to both Artemis Ward and Mark Twain

...common sense is as rare as genius.

Emerson (1844)

2.1 COGNITIVE PSYCHOLOGY AND HEURISTICS

The lifeblood of cognitive psychology is the nature and limitations of the mental shortcuts we use in our everyday lives. Remarkably efficient—although often illogical—in many situations (Gigerenzer, 2000, 2007; Gigerenzer, Todd, & The ABC group, 1999), such shortcuts can bias or degrade decision-making in other situations (Gintis, 2009; Piattelli-Palmarini, 1994). According to the big-mistake hypothesis, many of the mental adaptations inherited from our Pleistocene ancestors create boundaries to our rationality (Richardson, 2007; Richerson & Boyd, 2005) that require consideration to minimize errors in judgment while navigating in the modern world. Some cognitive mistakes are made by individuals in isolation, whereas others emerge only when individuals are, or envision themselves as, interacting members of a group. This chapter focuses primarily on the first type, those of the individual independent of a social context. Later chapters will explore the reasoning errors associated with groups.

The Nature and Use of Ecotoxicological Evidence. DOI: https://doi.org/10.1016/B978-0-12-809642-0.00002-9

Here is an example to start the reader along the path taken in this chapter. Rizak and Hrudey (2006) conducted an online survey of two groups of environmental specialists, asking them the following question:

> *[Hypothetical scenario] Monitoring evidence for a(n) [Australian] city has indicated that in treated drinking water, a pesticide, say 'atrazine', is truly present above the recognized standard method's detection limit once in every 1000 water samples from consumers' taps. The analytical test for the pesticide has the following characteristics:*
> - *95% of tests will be positive for detection when the contaminant is truly present above the detection limit, and*
> - *98% of the tests will be negative for detection when the contaminant is truly not present above the detection limit.*
>
> *Q[uestion]. With these characteristics, given a positive result (detection) on the analytical test for the pesticide in the drinking water system, how likely do you think this positive result is true?*
> *Provide either a probability estimate _____ or your scale of agreement.*
> ☐ *Almost certain (95 to 100%)*   ☐ *Very unlikely (5 to 20%)*
> ☐ *Very likely (80 to 95%)*   ☐ *Extremely unlikely (0 to 5%)*
> ☐ *More likely than not (50 to 80%)* ☐ *Don't know*
> ☐ *Less likely than not (20 to 50%)*

The two polled groups were Australian Water Association members and individuals listed in the Association of Environmental Engineering and Science Professors directory. The majority of the 352 responders had more than 10 years of experience and 42.3% were directly involved in evaluating water quality monitoring results intended to protect human health. Most choose "Almost certain" or "Very likely" despite the correct answer being "Extremely unlikely." Surprised at these results, I surveyed environmental science professionals and students in southern China (50 students in the 2009 advanced course, Eutrophication and Environmental Risk Assessment at Xiamen University), central China (26 graduate students in a 2010 practical environmental statistics course at Huazhong Normal University), Spain (174 attendees of the 2010 SETAC Seville International Meeting), southern India (57 attendees of the 2011 India Erudite Program lecture at Cochin University of Science and Technology), and the United States (37 attendees of a 2014 invited seminar at the University of Georgia's Savannah River Ecology Laboratory). Meta-analysis of the results (Fig. 2.1) indicates that only 10% of responders identified the correct answer, and consistent with the Rizak and Hrudey study, most chose "Almost certain" or "Very likely." Based on the numbering scale incorporated into the survey, the correct answer was <5% but most chose 80%–100% certainty. The inescapable conclusion is that trained environmental science professionals and students throughout the world misinterpret straightforward monitoring information in a consistent way.

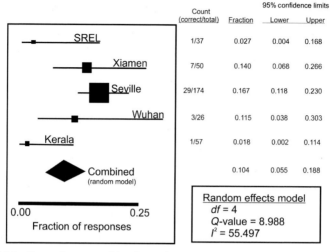

	Count (correct/total)	Fraction	95% confidence limits Lower	Upper
SREL	1/37	0.027	0.004	0.168
Xiamen	7/50	0.140	0.068	0.266
Seville	29/174	0.167	0.118	0.230
Wuhan	3/26	0.115	0.038	0.303
Kerala	1/57	0.018	0.002	0.114
Combined (random model)		0.104	0.055	0.188

0.00   0.25
Fraction of responses

Random effects model
$df = 4$
Q-value = 8.988
$I^2 = 55.497$

FIGURE 2.1 Forest plot summarizing the results (number selecting the correct answer and total number of answers) from five surveys. The *square symbol* indicates the fraction providing the correct answer with the symbol size reflecting the number of responses. The Count and Fraction columns reflect the number of correct responses/total number of responses and the fraction of all responses that were correct (e.g., SREL had 1 correct response of 37 responses or 0.027). A random effects model was applied during meta-analysis because the meta-analysis df of 4 was much smaller than the Q-value (Q reflects the relative magnitudes of the observed variation and the within-occasion error (Borenstein, Hedges, Higgins, & Rothstein, 2009)). The I^2 value indicates that roughly half of the variation was associated with differences among occasions. The *diamond symbol* displays the combined results with the left and right tips being at the lower and upper 95% confidence limits for each weighted estimate. SREL = University of Georgia's Savannah River Ecology Laboratory in 2014, Xiamen = Xiamen University in 2009, Seville = SETAC Seville International Meeting in 2010, Wuhan = Huazhong Normal University in 2010, and Kerala = Cochin University of Science and Technology in 2011.

A baffling incongruity emerges from this example. Most ecotoxicologists and environmental risk assessors would refuse to use results from an analytical method for measuring pollutant concentration if it were as biased as the above judgments. An analytical method that consistently reported very high concentrations (e.g., 80−100 µg/L) when, in fact, concentrations were quite low (e.g., >5 µg/L) is clearly unacceptable. Applying such a flawed analytical method would be indefensible if a better one were available. Why then are we comfortable with applying the informal judgments just described for making environmental hazard and risk decisions if they are as misleading as the hypothetical unacceptable analytical method just described, and better decision-making methods are available? Also, might there be other undetected judgments about chemical hazards—or even chemical presence—as biased as the above survey suggests? Answers to this last question and suggestions for reaching sounder judgments will be provided here and in Chapter 5, Individual Scientist: Reasoning by the Numbers.

2.2   ORIGINS AND NATURE OF EVERYDAY HEURISTICS

2.2.1   Fast and Frugal, But Potentially Biased, Heuristics

...mental economy... [refers to] certain routine mental calculations, those that come to us with the greatest ease and spontaneity and that seem irresistible that we barely know we possess, use, or abuse them.

Piattelli-Palmarini (1994)

Before delving into the cognitive compromises that might impact our science, it is only fair to give air to the assertion by Gigerenzer and his co-investigators that many mental shortcuts are remarkably fast, frugal, and still exceedingly useful (e.g., Gigerenzer & Brighton, 2009; Gigerenzer & Goldtein, 1996; Gigerenzer, 1991; Gigerenzer & Gaissmaier, 2011; Gigerenzer, 2008; Gigerenzer, et al., 1999). Gigerenzer reasons convincingly that these mental shortcuts are much more efficient in many everyday instances than more logically rigorous approaches. As an example, it would be a waste of time and effort to apply a rigorous decision-making algorithm to the task of picking people to invite to a dinner party or picking a restaurant for an outing. Nevertheless, these shortcuts can lead to substantial errors in judgment if they enter without notice into more complex decisions such as those involved in scientific studies and risk assessment. For this reason, their potential for creating error in judgments is highlighted below.

The errors described here emerge from our evolved mental tools for reasoning as individuals.

The formal term for the cognitive tools about to the explored is heuristic. A heuristic is a method, approach, or rule for efficiently making a judgment or decision that does not draw on rigorous, logical methods. In everyday language, a heuristic is a rule-of-thumb for making judgments. Gigerenzer and Gaissmaier (2011) provide a more explicit definition, "... a strategy that ignores part of the information, with the goal of making decisions more quickly, frugally, and/or accurately than more complex methods." It should be clear from this definition that many are applied automatically and without thinking. Heuristics evolved to allow an individual to (1) consider fewer cues or less evidence, (2) reduce the effort in retrieving cues, (3) simplify cue weighting during judgment, (4) integrate less information or evidence, or (5) examine fewer alternatives (Gigerenzer & Gaissmaier, 2011; Shah & Oppenheimer, 2008). Whether a particular cognitive trait became part of our adaptive toolbox (Gigerenzer et al., 1999) was dictated by whether it materially enhanced the fitness of early humans living in small hunter-gatherer groups on the Pleistocene plains of Africa. Many remain adaptive but some are maladaptive in certain modern situations.

A few heuristics are believed to emerge from subconscious rules of language. As an important example, Gigerenzer (2007) argues that the conjunction fallacy described in Section 2.2.2 is not a fallacy at all when the expectations of language economy are understood. A typical study of the conjunction fallacy begins with a subject being told a story like the following. Three good friends, Peter, Paul and Mary, go out to eat one night at a favorite restaurant and then return to their respective homes. Paul calls Mary the next morning, and during the course of their conversation, they realize that both had contracted food poisoning. Typically, the narrative ends at this point and the listener is asked if it is more likely that only they have food poisoning or Peter has food poisoning too. Subconsciously applying the maxim of relevance, the listener responds that likely all have food poisoning. But this is the wrong choice in the context of probability logic because the probability of two events (two friends with food poisoning) having occurred will always be higher than that of all three events (three friends being poisoned) occurring.

The maxim of relevance just mentioned is the instinctive rule minimizing superfluous information in a story and encouraging only relevant detail (Gigerenzer, 2007). Perhaps, the clearest example of misdirection based on the maxim of relevance comes from the DENSA webpage, http://home.utah.edu/~msm25/Funnies/densa.html. A question in their intelligence test is "Some months have 31 days; some months have 30 days. How many have 28?" The answer is "All twelve months have 28 days." In attempting to answer this trick question, an unsuspecting individual automatically disregards the irrelevant and banal context of "if all months have 28 or more days, do all months have 28 days?" Instead, they attempt to reframe it as the more relevant question, "Which month(s) has 28 days?" The question does not measure intelligence: it measures how much an individual tries to find relevance in the question.

What makes sense based on an innate language rule is nonsense in the context of probability logic. It is assumed that exaptation[1] of the maximum of relevance for purposes of probability reasoning creates this problem. Unfortunately, for the inattentive ecotoxicologist, the data-based narratives extracted from nature rarely obey the *maxim of relevance*, making the danger of the conjunction fallacy quite real.

2.2.2 Prominent Heuristics and Cognitive Biases

. . . the human mind resembles those uneven mirrors which impart their own properties to different objects, from which rays are emitted and distort and disfigure them.

Bacon (1620)

The important point here is that although evidence and reality constrain our beliefs, they do not do so completely.

Gilovich (1991)

---

1. Exaptation was defined originally by Stephan J. Gould as the repurposing of a trait that evolved for another purpose. For example, feathers evolved in dinosaurs for thermal regulation were repurposed later for flight in birds. See Gould (1991) and Buss, Haselton, Shackelford, Bleske, and Wakefield (1998) for discussion of adaptation, co-opting exaptation, and spandrels relative to evolutionary psychology.

Twenty-seven tendencies with the potential of compromising cognition by individuals, including scientists and risk assessors, are described below. Thoughtful reading will give the correct impression that many are intermeshed and perhaps become distinct only in the highly structured experiments of cognitive psychologists. Indeed, all are integrated items from the same cognitive toolbox. The first 18 relate to tendencies that could occur in individuals in isolation. The next three cognitive errors (2.2.1.19 to 2.2.2.21) are also made by individuals; however, they are preconditioned by real or imagined interactions with others. Errors 2.2.2.22 to 2.2.2.26 are made when dealing with probability. The last bias is not a cognitive bias per se or even linked to a heuristic, but it is relevant to the intent of this chapter. It involves perception errors resulting from how our brains process and attempt to make sense of visual information.

2.2.2.1   Overconfidence Bias

Generally, humans are more confident in their judgments than justified by objective evidence. The classic work of Fischhoff, Slovic, and Lichtenstein (1977) illustrates this point. Their study involved responses to several questions. The first was the open-ended question, "Absinthe is a _____", with two choices: a precious stone or a liqueur. Each response was scored relative to its accuracy and the responder's confidence in their answer. Confidence was expressed as how probable it was that they were correct. Responses scored with absolute certitude were correct only 30% of the time. The second question concerning the frequency of human deaths resulting from a variety of causes is more germane to the subject of this book. The questioner presented two causes of death in the United States, and the responder was asked to pick the most frequent cause of the two and to indicate their certainty in their choice. Confidence was expressed this time as odds with 1:1 indicating equal probabilities of being right or wrong, i.e., a guess. Within a confidence odds range of 3:1 to 100:1, the actual accuracy of responses did not change appreciably, suggesting again overconfidence in judgments. This and more recent studies conclude that humans are too confident in their judgments under most conditions.

It is easy to speculate from the last example that overconfidence might influence a risk assessor's judgments about which of several candidate stressors might have caused a harmful change in the biota of a contaminated site. Any assertion that overconfidence bias is rare in ecotoxicological research and ecological risk assessment activities would seem presumptuous. Yet, to my knowledge, there has been minimal effort expended in the ecotoxicology and ecological risk assessment literature to identify and then apply calibration curves (see Section 2.3.2) to compensate for this bias.

2.2.2.2   Availability (Ease of Recall) as Influenced by Vividness or Emotional Content

Our acceptance of an explanation can be influenced by how easily it comes to mind, that is, its availability for recall (Tversky & Kahneman, 1974). One way that availability manifests is through the impact of emotional vividness on recall. Piattelli-Palmarini (1994) explained this effect and followed it with an example, "The easier it is to imagine an event or a situation, and the more the occurrence impresses us emotionally, the more likely we are to think of it as also objectively frequent." The example involved opinions about whether fireworks accidents or diabetes is more of a lethal risk to US citizens. The easier visualization and higher emotional impression of a fireworks accident relative to the incrementally debilitating effects from diabetes mellitus drew many people to the wrong conclusion that fireworks accidents are more of a risk.

It is prudent to reflect on how much this bias might also influence the practice of ecotoxicology. Perhaps, this bias reinforces ecotoxicologists' and risk assessors' continued preoccupation with acute mortality to the neglect of population effects of pollutants. The image of a fish kill is more emotionally vivid than the slow, but equally impactful, disappearance of a species population over several years due to chronic toxicant exposure.

2.2.2.3   Availability (Ease of Recall) as Influenced by Social Accentuation

This foible involves the tendency to more readily recall a fact, explanation, or hypothesis if it was communicated in the context of people. We, as social organisms, find a story more salient if a message or information was delivered through social actors instead of straightforward statements of facts (Gilovich, 1991; Rabin, 2003). For example, *The New Yorker* published an engaging article about the harmfulness of the herbicide, atrazine, by focusing on one researcher (Aviv, 2014). Tyron Hayes's role in this story was set by the article's opening line, "After Tyrone Hayes said that a chemical was harmful, its maker pursued him." This narrative of a scientist's struggle to overcome daunting odds was much more impactful than an objective review of the growing body of atrazine studies suggesting harmful effects of atrazine, such as the meta-analysis by Rohr and McCoy (2010).

Similarly, how a person feels about the individual conveying information also biases their willingness to accept the information. Does the individual have the same training and frequent the same professional meetings that I do? Are they employed by a university, regulated industry, or regulatory agency? These influences will be described in more detail in Chapters 6−8.

2.2.2.4   Availability (Ease of Recall) as Influenced by Familiarity

Availability is also at the core of the familiarity heuristic (Tversky & Kahneman, 1974). A familiar explanation tends to be favored over an unfamiliar one even if evidence provides more support for the unfamiliar one. Preference might be so strong that the familiar explanation is chosen with only token effort being expended to identify others and compare all plausible explanations.

The recognition heuristic is an interesting twist of this phenomenon that gives insight about why this bias emerged in human reasoning. Gigerenzer (2008) explains the everyday advantage it provides to anyone with so little direct ("local") knowledge that they must guess when deciding. "The recognition heuristic is successful when ignorance is systemic rather than random, that is, when recognition is strongly correlated with the criterion" (Gigerenzer, 2008). He gives the example of guessing whether Detroit or Milwaukee had the most inhabitants. An American with considerable local knowledge would likely guess correctly that Detroit was the largest. A German with much less local knowledge might subconsciously rely on recognition: Detroit is a city that they have heard much more about than Milwaukee, so the best guess is Detroit. The recognition heuristic is not strictly logical but the correlation between name recognition and population size must do if one is ignorant of specifics. A more relevant example might be ecotoxicologist's answers to the question, "Is a Curie or Gray a measure of absorbed radiation dose?" Those who have worked with radioactive contaminants would likely have enough local knowledge to correctly pick the Gray. Based on recognition alone, those without much radiation training would tend to guess the Curie. This heuristic provides an advantage in situations of high uncertainty (i.e., little local knowledge) but carries much too high of an error risk for inference to use in scientific research and risk assessment. The problem for the unwary decision maker is that the shift from relying on local knowledge to the recognition heuristic occurs automatically, making it easy to unknowingly make a decision based on recognition.

The familiarity bias has been demonstrated to influence medical diagnostics (Croskerry, 2003; Dawson & Arkes, 1987), so it is very easy to imagine it influencing judgments of a similar nature made by risk assessors. An ecological risk assessor familiar with the effects of chemical contaminants could easily be biased toward identifying chemicals as the causative agents for diminished biodiversity at a highly modified Superfund site and paying much less attention to co-occurring physical habitat alteration and fragmentation. The assessor would be prone to this bias in the typical case where a decision must be made in the presence of high uncertainty, again that being, minimal local knowledge.

2.2.2.5   Confirmation Bias

Wason (1960) demonstrated the human tendency to gather evidence to confirm hypotheses rather than objectively weighing the merits of all plausible hypotheses. Twenty-nine psychology students were given the sequence of three numbers, 2, 4, 6, and told that the sequence was generated with a simple rule that the examiner had in mind. To discover the undisclosed rule, each student was to form a hypothesis about what the rule was and then write down another set of numbers conforming to the hypothesized rule. The examiner would then state whether the new sequence did or did not conform to their rule. If the student's sequence was wrong, they were asked to reevaluate their hypothesis and produce another sequence that they believe now conforms to the examiner's rule. This new sequence was judged again by the examiner and the process continued until the individual student was confident enough that they had discovered the rule behind the original sequence. There was a marked resistance of students who had made two or more incorrect conclusions to question their own hypotheses. The tendency was to try to confirm a previous hypothesis despite contrary evidence. Mynatt, Doherty, and Tweney (1977) performed a more complex experiment more like scientific inquiry than Wason's experiment and also found a tendency to confirm hypotheses and to avoid falsification unless evidence was very strong against the hypothesis. Confirmation bias and a lack of skepticism about their own favored hypotheses by scientists were recently documented by Strickland and Mercier (2014). This confirmation bias in the context of favored scientific hypotheses will be discussed further in Chapter 3, Human Reasoning: Within Scientific Traditions, and Rules and Chapter 4, Pathological Reasoning Within Sciences.

2.2.2.6 Anchoring

Similar to confirmation bias, anchoring is the overweighting of something that comes before another in estimation or cognitive evaluation. Tversky and Kahneman (1974) gave the example of two groups of high-school students asked to quickly estimate the product of a series of numbers. One group was given the sequence, $8 \times 7 \times 6 \times 5 \times 4 \times 3 \times 2 \times 1$ and the other was given $1 \times 2 \times 3 \times 4 \times 5 \times 6 \times 7 \times 8$. The median answer from both groups should be 40,320 if the commutative law of multiplication was the only rule in play. The observed median answers for the descending and ascending sequences were 2250 and 512, respectively. What appears to be happening is that students do the first few computations and then tend to quickly do estimations for the remaining numbers. Students' answers were biased toward the first numbers in the sequence. Subconsciously, people tend to also anchor when informally estimating the probabilities of different events occurring, relative plausibilities of competing hypotheses, or relevance of evidence in decision-making (Piattelli-Palmarini, 1994; Tversky & Kahneman, 1974).

Anchoring can be defined in a precise way more applicable to cognitive errors that ecotoxicologists or ecological risk assessors might make. Anchoring is a dependency of belief on initial conditions. This might take the form of a preference for one option or approach that emerges during the first steps of an investigation or assessment. For instance, a strong argument could be made that our continued reliance on the LC50, *a measure of relative toxicity*, as a predictor of *ecological consequences* is a case of anchoring. This metric was borrowed from mammalian toxicologists at the beginning of our efforts to predict pollutant effects in the environment. Despite innumerable sound arguments for discontinuing such use, anchoring and the large amount of existing LC50 data continue to shackle us to this passé metric.

In my opinion, anchoring also manifested in the Aquatic Dialogue Group (1994) tiered pesticide risk assessment scheme. On page 15 of this report, the purpose of the chosen approach is stated: "The process of [pesticide] risk assessment should be tiered to reduce the need for unnecessary assessments and delays in the regulatory process." The first tier used an estimated environmental concentration (EEC) of the pesticide and the LC5 for the most sensitive of a set of test species. No further assessment was needed if the $EEC < LC5_{sensitive}$, and *if an NOEC was available,* $EEC < NOEC_{sensitive}$. The first tier cheaply determines if a pesticide presents no significant risk and "no further assessment [is] necessary." They reasoned that there was no need to go to a higher and more costly tier if the worst case was not uncovered in this tier. However, if it does not pass Tier 1, a Tier 2 probabilistic assessment is begun in which distributions of ECC values are compared to distributions of species sensitivities (such as LC5 values). The assessment stops here if the curves indicate that no more than 10% of species are potentially affected. If not, the process continues with more explicit details being added until the conditions of the assessment are satisfied.

Notionally anchored by approaches taken in the early history of ecotoxicology, nowhere in the process are data for population or community effects data required. Illogically, if the worst case of mortality to individual organisms is not realized in Tier 1, the Aquatic Dialogue Group assumes that more subtle effects are improbable. The assumption seems to be that the population and community levels are protected if the individual level is protected based on toxicity to a collection of test species. The decision that the community will be safe if fewer than 5% or 10% of the individuals for tested species are not affected, and any affected species is not an "endangered species or organisms of high ecological, commercial, or recreational significance" seems more conventional than thoughtful. Ecotoxicology emerged during a time when instances of blatant effects to individuals were common and, accordingly, initial methods focused on effects to individuals. (Early ecotoxicologists, most notably John Cairns, Jr., were unable to broaden testing to include higher levels of biological organization (e.g., Cairns & Pratt, 1993; Cairns, 1983, 1986, 1992)). Evidence collected by and since the time of the Aquatic Dialogue Group report suggests that individual effect metrics are unreliable indicators of population effects (Calow, Sibly, & Forbes,

1997; Forbes & Calow, 2015; Forbes, Calow, & Sibly, 2001, 2008; Forbes et al., 2011). Nor would they be adequate indicators for community effects.

The early dichlorodiphenyltrichloroethane (DDT) impact to many raptor and piscivorous species would not have been predicted by this tiered approach because the community processes of trophic exchange and magnification were key to impactful DDT exposure. Nor would the current plummeting oriental white-backed vulture populations (Balmford, 2013) have been prevented if this tiered protocol had been applied for the veterinary drug, diclofenac. Due to anchoring of protocols to organismal effect metrics for a subset of test species, the critical ecological interactions resulting in vulture poisoning while scavenging cattle carcasses would have been overlooked. The emphasis on individual effects data is unjustified and the tiered pesticide assessment scheme seems primarily a product of scientific anchoring and regulatory expediency.

2.2.2.7 Framing Error

This error arises if the way a question or situation is presented influences an individual's response. McNeil, Pauker, Sox, and Tversky (1982) split medical doctors into two groups, stating to one that a particular surgical procedure has a mortality rate of 7% within 5-year post-surgery and to the other group that the surgical procedure has a survival rate of 93% within 5 years. The doctors were then to decide based on these identical, but differently framed, facts whether they would recommend the procedure to one of their patients. Physicians presented with the statistics framed in terms of survival were more likely to recommend the surgery than those presented with statistics framed in terms of mortality.

Schuldt, Konrath, and Schwarz (2011) detected this bias more recently in judgments about the environment. They questioned US laypersons about their belief in the phenomena using either the phrase "global warming" or "climate change." Republicans decided that the phenomena was real 44.0% of the time if "global warming" was the wording, but 60.2% of the time if "climate change" was the phrase. Democrats were less influenced by wording (86.9% for "global warming" and 86.4% for "climate change"). Clearly, question framing is important if one wishes to persuade voting citizens of the seriousness of climate change.

Another example (Viscusi & Chesson, 1999) illustrates framing-related bias in decision-making. Two hundred and sixty-six coastal business owners or managers who could potentially suffer losses due to climate change were presented with risk estimates reflecting different degrees of ambiguity. Their decisions shifted as loss probabilities changed. "People exhibit 'fear' effects of ambiguity for small probabilities of suffering a loss and 'hope' effects for large probabilities" (Viscusi & Chesson, 1999). Ambiguity-dependent fear and hope shifted decisions away from those of a hypothetical rationale decision maker. Supporting the emotion-based interpretation of the Viscusi and Chesson study, De Martino, Kumaran, Seymour, and Dolan (2006) linked framing during decision-making to increased activity in the amygdala, the part of the brain associated with emotions, emotional behaviors, and motivation.

It takes little effort to envision similar shifts in judgment during deliberations by ecotoxicologists and ecological risk assessors. For instance, would the acceptability of consequences be judged differently from the species sensitivity distribution approach if they were expressed as 5% of species would be lost instead of 95% species would be protected?

2.2.2.8 Acquiescence

This is the tendency to accept a problem as initially presented. Not so for anchoring, acquiescence seems to emerge from our innate tendency to be socially agreeable and to avoid dissonance. As a common example, acquiescence often enters unexpectedly into "Agree/Disagree" surveys that use the ratings system: agree completely, agree somewhat, neither agree nor disagree, disagree somewhat, or disagree completely. Saris, Revilla, Krosnick, and Shaeffer (2010) indicated that responders are prone toward agreement in these surveys regardless of the question content due to engrained habits of congeniality and aversion to social conflict. Gilovich (1991) gives another example in which subjects were told that either introversion or extraversion were associated with good scholastic achievement. Responders tended to indicate that they were more introverted or extroverted depending on which quality was presented as positively related to good scholastic performance. Relative to evidence to support their claims, responders recalled more personal instances of their displaying the positive achievement-related quality than of the other quality.

Acquiescence can be cast in a more general context to mean "when we are faced with a reasonable formulation of a problem involving choice, we accept it in terms in which it is formulated and do not seek an alternative form" (Piattelli-Palmarini, 1994). When acquiescence is prominent, the way the problem is presented determines how it is thought about during deliberations and decision-making.

As a result of there being both technical and scientific goals in the ecotoxicology, there are numerous opportunities for acquiescence to enter into judgments. Scientific activities in ecotoxicology aim to organize knowledge about contaminants based in explanatory principles, whereas the technological goal of ecotoxicology is to develop and apply tools for acquiring a better understanding of contaminant fate and effects in the biosphere (Newman, 2015). In addition to creating tools for scientific research, technical studies might create tools for practical endeavors to document or solve specific problems such as determining if a legal violation has occurred in a particular situation. Scientific studies traditionally apply hypothetico-deductive inquiry, often including formal falsification, working hypotheses, and statistical testing. In contrast, a technical endeavor to develop a standard method or approach for regulatory purposes inevitably will require a validation process that tries to demonstrate utility of the method. Acquiescence can bias a validation exercise through the desire to support a newly introduced regulatory approach rather than reject it from further development. Individuals invest time and stake their professional standing when proposing a method, so others can be swayed into acquiescence by the involuntary desire to be collegial and professionally civil. In my opinion, this occurs often enough to deserve special consideration during validation exercises.

2.2.2.9   Segregation/Isolation Effect

Segregation occurs when a problem or situation is addressed with inadequate consideration of its more global context. Or, in some instances, when the choice among alternatives is simplified by disregarding common features of alternatives and focusing only on differences. A problem becomes more tractable and less stressful during decision-making by decreasing the number of features and alternatives (Kahneman & Tversky, 1979). Piattelli-Palmarini (1994) remarked that the framing heuristic described above has components of both acquiescence and segregation. Acquiescence biases us to accept the presented context and segregation tends to isolate the problem from its more global context, including alternative features, explanations, or solutions. The evolved motive here is to simplify the process of deciding (Kahneman & Tversky, 1979).

Segregation is expedient for many commonplace decisions, but it can produce substantial error in complex decisions such as those of ecotoxicologists and risk assessors. The danger would be especially high when co-occurring with substantial acquiescence. For example, a working committee might be called together by a regulatory agency to consider the merits of including a promising method in a revised methods manual for ecological risk assessment. The regulatory agency representative might charge the committee with evaluating the method for the intended use in risk assessment. Implicit in this typical charge would the exclusion or subordination of other, potentially superior methods for accomplishing the same goal. Acquiescence and segregation could result in a suboptimal procedure being adopted. Much unnecessary effort would then have to be exerted in the future to dislodge the established method and replace it with a better one.

2.2.2.10   Time Inconsistency in Choices

Decision-making should involve a strategy in which the expected benefit, costs, and risks for each alternative are integrated to arrive at an optimal (maximized) decision (Janis & Mann, 1977). That optimal decision would be insensitive to time. In reality, people balance risks and costs differently if their decision involves immediate or delayed consequences (Frederick et al., 2002). Gintis (2009) sums up this inconsistency, "in the sense that if there is a long time period between choosing and experiencing the costs or benefits of the choice, individuals can choose wisely, but when costs and benefits are immediate, people make poor choices, long run payoffs being sacrificed in favor of immediate payoffs." Although the most logical choice would maximize utility, there is an increased inclination to take the benefit and discount costs with a shortening of the time period before the cost comes due.[2] A person offered either $1000 now or $5000 in 5 years is prone to take the immediate but less lucrative option. Yet, most people would wait if the higher $5000 offer was for 1 day, not 5 years, later. The tendency to favor immediate benefit also rises as the uncertainty grows about the long-term consequences of a decision (Cyert & March, 1963).

A habit might develop in an ecotoxicologist's research career of doing and publishing short, modestly informative studies instead of focusing on protracted and more impactful studies. Biochemical markers of response—but perhaps not plainly of harm—to individual organisms might dominate a prudent research career. Protracted, but potentially more rewarding relative to useful insight, studies of toxicant impact to an entire biochemical network or to ecological community interactions might be considered too time consuming or risky to venture. In risk assessment, a 1-year intensive study of avian biodiversity might be favored over a more informative and realistic one that included variation among five consecutive annual surveys. Relative to applying knowledge to decide how to remediate a contaminated site, an immediate removal of contaminated soil or sediment might be favored over a less costly decision to allow natural processes to bring concentrations to acceptable levels in a decade or two.

2.2.2.11   Value Centrality Bias

This error arises from the excessive influence of one's core values on decision-making. It was added to the usual list from the cognitive psychology and decision-making literature because it is common in applied sciences like ecotoxicology. The moniker, value centrality, was borrowed from Verplanken and Holland (2002) where it referred to values pivotal to how people define themselves. This bias arises if an individual asserts personal or institutional core values into judgments in a way that produces an objectively suboptimal judgment or

---

2. Briefly, utility analysis allows the calculation of probable benefit per unit of expenditure. As a simple example involving expected value, a lottery ticket might cost $2.00 and have a 1 in 5,000,000 likelihood of yielding $1,000,000 in winnings. The expected value of the lottery ticket is $1,000,000/5,000,000 or $0.20. Preoccupied by winning $1,000,000, a lottery player would spend $2.00 for a ticket with an estimated value of only $0.20. That being said, cost and benefits can be nonmonetary. Cost-utility estimations incorporating more than one simple attribute, for example, ticket cost plus the enjoyment of playing the lottery, become more complicated and prone to error if done intuitively. Pages 8–11 of Gintis (2009) provide a formal utility analysis that explains the time inconsistency phenomena in this context.

decision. The error can be exacerbated by an erroneous overestimation of the degree to which others share one's beliefs or attributes (Gilovich, 1991).

Examples of this bias can be easily found in the general and scientific literature. The influence of political party affiliation on belief in global warming (Schuldt et al., 2011) has already been described. An historical example can be found in Popper's (1956) discourse about scientific induction, "... Galileo wrongly rejected the lunar theory of the tides because ... it was part and parcel of the astrological theory of 'stellar influences', which he rightly felt was in bad taste." Ronald Fisher, arguably the father of modern statistics but indisputably an avid pipe smoker, wrote prolifically about the wrong-headedness of researchers claiming that smoking harmed one's health (e.g., Fisher, 1957, 1958a−c). (Sir Austin Barnard Hill's publications about the harmfulness of smoking were the primary target of Fisher's attacks. Hill will come up again in later chapters during discussions of criteria for gauging causal association.)

Value centrality-based errors in ecotoxicological decisions are likely to occur because of the multidisciplinary nature of the field, and the distinct scientific, technical, and practical goals of its practitioners. For example, identifying strongly with the environmental movement can sway judgment enough to produce a suboptimal decision. The influence of environmentalism on many individuals' core values is underscored by Nelson (2004) who argued that environmentalism has many qualities in common with religion. As such, tenets of environmentalism become central values making up many individual's self-identity. At the other end of the spectrum, individuals with satisfying careers in a regulated industry can hold beliefs that sway their judgments about environmental issues in the opposite direction. During natural resource damage assessment, a regulatory agency under obligation to specific laws, such as the US Fish and Wildlife Service's obligations under the Migratory Birds Treaty Act, might focus on the easily provable loss ("taking") of individual organisms while giving short shrift to more profound, but more difficult to legally prove, damage to ecological populations.

2.2.2.12 Compulsion to Find Order, Patterns, and Relationships

Modern humans inherited an innate urge to find patterns and explanation, even where none exists. The advantage of this compulsion is self-evident when patterns, trends, or easy explanations do exist. When a nonrandom pattern or valid explanation is not discernible with available evidence, the urge to assign one can produce errors in judgment.[3]

---

3. In Stephan J. Gould's aforementioned work describing exaptations (Gould, 1991), he also proposed another term, spandrels as pertinent to evolution. Spandrels are "presently useful characteristics [that] did not arise as adaptations ... but owe their origins to side consequences of other features" (Gould, 1991). Modern human's urge to find patterns in data can be understood as being a spandrel.

The clustering illusion occurs when one incorrectly perceives nonrandomness in a random pattern (Fig. 2.2). An individual succumbing to this illusion might examine a sequence of outcomes from fair coin tosses and judge the number of clusters of sequential heads outcomes to be suspiciously more common than expected from chance alone. Two factors contribute to the clustering illusion. First, the observer fails to understand that sequences generated by random processes will include clusters. Having misjudged the sequence as being suspiciously nonrandom, the observer then falls victim to the compulsion to find order in the sequence, producing the clustering illusion.

The urge to find order also manifests with less abstract processes. As an example, basketball fans often say a player has "hot hands" when they make a sequence of baskets during a game. Statistician, R. Kass and his student, K. Hsiao (see Kass & Raftery, 1995) analyzed Larry Bird's 1986–87 baskets, finding no evidence of nonrandom streaks of hot handedness. This was not very convincing because Bird was a reputed "steady shooter," so they found a player reputed to be a "streak shooter" (Vinnie Johnson, 1985–89

FIGURE 2.2   Do large clusters of black beads seem too common to have come from a random stringing process? It is difficult to be intentionally random, and perhaps as a consequence, you might guess that the author failed to produce a random series. A formal χ^2 test was performed to determine if there is evidence in the results for deviation from the expected Poisson distribution of black bead cluster sizes ($\chi^2 = 4.401$, $df = 6$). According to current statistical convention (but see Chapter 5: Individual Scientist: Reasoning by the Numbers), there is a probability of 0.62 that these or more extreme results occurred by chance alone: there is little evidence for rejecting the hypothesis that stringing of beads was random.

seasons) for comparison. Again, they found no evidence of hot handedness. Clustering was being attributed incorrectly to successful basketball shots.

The urge to find patterns also appears in simple data plots, so judgments from plots and regression methods require diligence to avoid seeing a pattern or trend where none exists. Matthews (2000) illustrated the dangers of regression-based identification of dubious trends by drawing on the old German explanation given to inquisitive children that storks deliver babies. The number of babies born in 17 European countries was compared to the number of white storks nesting in each country (1980−90). There was a statistically significant, positive regression slope, suggesting that storks do deliver babies! In an equally whimsical example, Messerli (2012) reported in the prestigious *New England Journal of Medicine* that regression analysis revealed a significant, positive relationship between a country's annual per capita chocolate consumption rate and the number of Nobel Prizes awarded to its citizens.[4] The simultaneously obvious and dubious conclusion being that eating chocolate makes you really smart. These examples highlight in an amusing manner the pervasive problem with regression analysis that people "often invent spurious causal mechanisms" (Tversky & Kahneman, 1974).

As a related aside, Tversky and Kahneman (1974) highlight another common cognitive bias emerging during regression exercises, that being a failure to understand the more subtle concept of regression toward the mean. The classic example of regression toward the mean is Galton's analysis of heights of fathers and sons. Heights of sons of tall fathers and of short fathers tend toward the regression line that predicts mean height of sons for specific height fathers. Galton referred to this as regression [of sons' heights] back toward the mean, providing the general approach with its obscure name. People tend to overlook the regression toward the mean aspect of the results and focus solely on explanations for relationships, such as, "X causes Y." With the example here, tall fathers do tend to have tall sons but because father's height is not the sole contributor to a son's height, regression to the mean is also occurring. Tversky and Kahneman conclude that individuals are poorly equipped cognitively to fully interpret regression results.

There are so many opportunities for these kinds of errors in ecotoxicology and ecological risk assessment activities that discussion of specific examples would generate very little additional insight. The point to keep in mind is that humans are inherently prone to seeing clusters and trends, or providing *ad hoc* explanations where none are justified by evidence.

---

4. There is the possibility of a value centrality bias here because Doctor Messerli reports in the disclosure section of his paper that he is prone to "regular daily chocolate consumption, mostly but not exclusively in the form of Lindt's dark varieties."

2.2.2.13 Differences for One- and Two-Sided Events

A two-sided event is one in which both outcomes are equally impactful or memorable to the decision maker. As an illustration, equally unmemorable would be two alternative outcomes in which a person neither risks nor gains much either way. In contrast, a one-sided event is one in which one outcome is more memorable or has more emotional impact than the other. An outcome is very easily recalled if it supports one's expectations or represents high risk. Outcomes from one-sided events are more easily retrieved during evidence weighing than those from two-sided events.

One particularly relevant one-sided event is a confirmation event (Gilovich, 1991). An event might be especially memorable in one's study if it supports prestudy beliefs or a favored hypothesis. An outcome that was nonconfirmatory would not be as easily recollected. This difference in one-sided event outcomes is especially insidious during scientific judgments. Dewey (1910) discusses this problem by relating Charles Darwin's personal antidote for avoiding it, "Darwin remarked that so easy is it to pass over cases that oppose a favorite generalization, that he made it a habit not merely to hunt for contrary instances, but also to write down any exception he noted or thought of—as otherwise it was almost sure to be forgotten."

As discussed in detail in Chapter 5, Individual Scientist: Reasoning by the Numbers, the pervasive misapplication of null hypothesis statistical testing creates the risk of such overemphasis on significant relative to nonsignificant outcomes. Another ecotoxicology example might be the unequal recall of studies supporting or not supporting one's favored explanation of evidence. Darwin's habit seems worthy of more consideration by ecotoxicologists and ecological risk assessors.

2.2.2.14 Endowment and Mere Ownership Effects

Described first in financial investing, the endowment effect (or divestiture aversion) is the reluctance to abandon a failing investment despite accumulating evidence of failure. The root is an aversion to taking a loss. The loss of some money weighs more heavily in a decision than the profit that could be gained by moving any remaining funds into a better investment. As counterintuitive as it may seem, "people appear to be about twice as averse to taking losses as to enjoying an equal level of gains" (Gintis, 2009). Kahneman and Tversky (1979) refer to this as a failure to make peace with losses.

This effect is seen in many forms of decision-making. A scientist(s) might remain committed to an explanation or causal hypothesis despite the accumulation of contrary evidence. In this context, the mere ownership effect, which is a manifestation of this same general foible, might be the most explicit description of what is occurring. The mere ownership effect

(Beggan, 1992) is the tendency to value something more than objectively justified simply because one owns it. The endowment or mere ownership effect in the case of scientific theory might be referred to as theory tenacity. Investments can also be made in procedures and technologies that, when a better one appears, we are reluctant to abandon.

There is no reason to believe that endowment or mere ownership effects do not compromise decisions by environmental scientists or assessors (Newman & Evans, 2002). In my opinion, the continued overreliance on LC50 and related toxicity metrics for individuals is a blatant example of the endowment effect.

2.2.2.15   Status Quo Bias

This resistance to change is broader than the endowment effect in that a loss or sense of ownership need not be involved. Samuelson and Zeckhauser (1988) begin their description of this bias by noting that all decisions have some risk and uncertainty associated with them. The decision to stay with the *status quo* can involve less emotional stress or anxiety than one to move away from the *status quo*. They illustrated this bias with a questionnaire-based study in which framing of alternatives impacted the likelihood of choices. Neutral framing expressed all alternatives as having equal footing and *status quo* framing as having one alternative that was the status quo. The results demonstrated a clear *status quo* bias; however, its strength varied. The variation depended on the degree to which an individual favored a particular option. So*, a status quo* bias exists in decision-making but varies in complex ways.

The continued reliance on LCx metrics at a fixed duration—often 96 hours—is a good example of *status quo* bias in ecotoxicology. It is unlikely that this set duration would be chosen if a panel of experts were asked using neutral framing to choose from two candidates the one most useful for predicting lethal consequences of toxicant releases: one that measured the influence of exposure intensity (concentration) at a set duration or one that simultaneously included duration and intensity of exposure. Most panelists would understand that toxicant releases vary in both intensity and duration, and as a consequence, a metric that accommodated both would be the best. Numerous books detail methods for producing models including both duration and intensity of exposure (e.g., Cox & Oakes, 1984; Miller, 1981; Crane, Newman, Chapman, & Fenlon, 2001; Marubini & Valsecchi, 1995). Newman and colleagues demonstrated their application in ecotoxicology (Newman & Clements, 2008; Newman, 2013, 2015) and risk assessment (Crane et al., 2001). Similarly, No and Lowest Observed Effect Concentration (NOEC and LOEC) metrics continue to be generated and used in ecotoxicology judgments despite a long series of publications criticizing them (see Newman & Clements, 2008; Newman, 2013, 2015).

Notwithstanding years of criticism of fixed duration LCx and NOEC/ LOEC metrics, regulatory documents such as those of the US EPA and OECD continue to give details for conducting tests designed for their estimation. When reading studies reporting these metrics, the poem, *Science*, by Alson Hawthorne Deming (1989) often comes to my mind. The poem describes a disturbingly incongruous science project exhibited in a high-school cafeteria in which mice were weighed before and after being chloroformed to death in an attempt to estimate the weight of the soul. A profound incongruity exists between the experimental intent and what is actually measured.

2.2.2.16 Risk Aversion

This aversion contributes to some biases including the endowment effect and *status quo* bias just discussed. An argument can be forwarded that risk aversion in itself is not always a misguiding heuristic. Nonetheless, risk aversion is included here to expose the reader to its influences on judgments.

Risk aversion can be a positive or negative force in decision-making. Anxiety aroused by the possibility of post-decision regret can increase vigilance and improve the chances of a thoughtful decision. However, excessive aversion or preoccupation with risk can color judgments and result in a sub-optimal decision. Under some conditions, risk aversion can even produce defensive avoidance or bolstering behavior. Janis and Mann (1977) indicate that if an individual believes that no more evidence is likely to be gathered by a specified decision deadline, they will pick the least objectionable of alternatives and then bolster that decision with ad hoc justifications. Bolstering can occur before and after a decision to reduce pre- and post-decision stress. If an individual believes that more information might be available in the near future, there is a lower likelihood of bolstering to support a selected alternative and a greater chance of an even-handed balancing of the alternatives. Janis and Mann (1977) refer to these differences in decision-making based on the likelihood of more evidence emerging as the expected information hypothesis.

Other influences on risk aversion exist such as wealth. Gintis (2009) applies utility computations to model the lessening of risk aversion as a person becomes wealthier. To a point, how willing an individual is to take financial risks increases as wealth accumulates. This bias is so predictable within a certain range of wealth that it can be quantified with an Arrow-Pratt coefficient.

It takes very little imagination to speculate that scientists will act similarly as a function of their accumulated stature (wealth) in their fields. The influence of wealth on risk aversion explains the outwardly counterintuitive finding of Borjas and Doran (2013) that exceptional mathematicians experience a drop in productivity after they are awarded the prestigious Fields

Medal. (This award is given to highly accomplished, young mathematicians to encourage more excellent achievements, and also to encourage others to similar levels of excellence.) The recipients "begin to play the field, studying unfamiliar topics at the expense of writing papers" (Bojas & Doran, 2013). This makes sense when the receiving of this award is envisioned as a large increase in the awardee's prestige—their intellectual wealth. They adjust their willingness to risk exploring other fields in the hopes of bringing innovative approaches or solutions back into their discipline. They can "afford" a consequent drop in productivity while exhibiting high-risk professional behavior.

Speculating from the above concepts, well-established ecotoxicologists uninfluenced by monetary or intellectual wealth might be more prone to propose risky, novel paradigms or hypotheses than are young ecotoxicologists who have yet to accumulate sufficient stature.

2.2.2.17 Risk Thermostat/Compensation

Risk perception is subjective and differences in perception produce differences in the willingness to accept risk. As one illustration of subjective perception of health-related risks, people commonly rate their own personal risk lower than that of others in the general population (Weinstein, 1989).

Adams (1995) applies the term, risk thermostat, to the differences among individuals in their willingness to accept risk and the tendency to adjust their risk level during actions or decisions accordingly. Two commonplace instances of risk adjustment are illustrative: the change in behavior associated with using automobile safety belts (Adams, 1995) and wearing helmets while bicycling (Gamble & Walker, 2016). Adams (1995) studied the influence of seat belt laws and corresponding changes in automobile accident fatalities in various countries. Contrary to popular belief, the evidence showed that mandatory use of seat belts did not save lives. They do reduce the risk *of the belt user* dying during an accident; however, individuals using seat belts feel safer and tend to drive less cautiously as a result. Belted driver's adjustment of their risk thermostats led to more outside-of-the-vehicle deaths. "The evidence with respect to seat belts suggests that the [British] law had no effect on total deaths, but was associated with a redistribution of danger from the car occupants to pedestrians and cyclists" (Adams, 1995). Using the term, risk-compensation, instead of risk thermostat, Gamble and Walker (2016) found that cyclists also unconsciously increase their risky behaviors when wearing safety helmets.

Risk taking in ecotoxicology decision-making involves risk to one's reputation or standing when reporting study results in the peer-reviewed literature or at professional meetings. Risk for an ecological risk assessor involves the correctness of conclusions in their written reports plus that of failing to solve the problem in hand. An added risk for a self-employed ecological risk assessor might be lost income associated with any diminished credibility. According to the risk thermostat concept, it is expected that some individuals will err more than the others in the caution applied in their ecotoxicological research and risk assessment conclusions and judgments. Adding co-authors or risk assessor team members to support conclusions might lessen an individual's risk by spreading the blame should something go wrong. Like the belted drivers or helmeted cyclists, an ecotoxicologist with co-authors or a risk assessor with team members might be inclined to display more risky decision-making. This specific feature of risk thermostat adjustment will be discussed again as groupthink in Chapter 6, Social Processing of Evidence: Commonplace Dynamics and Foibles.

2.2.2.18   Argumentum ad Ignorantiam

Walton (1996) provides a detailed discussion of errors associated with arguments from ignorance or using the Latin phrase, *argumentum ad ignorantiam*. This line of thinking takes the form of "No one has ever proven that flying saucers do not exist. Since you cannot disprove it, you must consider true—or at least possible—my claim of their existence." Presenting an argument from ignorance with a fringe topic such as flying saucers shows its shortcomings immediately; however, these shortcomings are not as obvious when a less extreme example is used. Perhaps, a more subtle, but equally fallacious, argument might be made during the registration process for pesticide X that acute toxicity test results are adequate for assessing pollutant X because no one has yet shown that they are inadequate evidence for this class of pesticide.

The precautionary principle has been proposed as a guiding tenet for ecological risk management. This conservative principle states that even in the absence of any clear evidence and in the presence of high scientific uncertainty, regulatory action should be taken if there is any reason to suspect that a contaminant might cause harm if released into the environment (Newman, 2015). The precautionary principle appears at first glance to be a subtle, but still logically flawed, argument from ignorance. Why does this particular example now seem more tenable? Walton (1996) believes that *argumentum ad ignorantiam* is not flawed if it is used in a decision process or argument to establish a presumptive or inconclusive position based on plausibility *that carries the implicit obligation to gather further evidence to reduce uncertainty in a final judgment.*

Argumentum ad ignorantiam alone is universally viewed as flawed during debate and decision-making. But, it can be useful as a starting point in decision-making if the obligation to gather further evidence is kept at the forefront of thinking and acted on in a manner that influences a final decision in the foreseeable future.

2.2.2.19   Self-Handicapping

At the root of self-handicapping is self-image protection. An individual might manipulate circumstances surrounding their behavior to preserve their self-image of being a worthy individual (Jones & Berglas, 1978). Naturally, the intensity of self-handicapping depends on an individual's self-esteem (Spalding & Hardin, 1999). Handicapping can take many forms, including some outwardly maladaptive behaviors. Substance abuse might be used as an extreme self-esteem preserving tool, becoming the reason for underachievement rather than the individual simply lacking the qualities necessary for success (Gilovich, 1991). Self-handicapping can also take the form of goal avoidance. If a task were completed, the results could imply poor performance; therefore, goal avoidance will assure self-image preservation. Self-handicapping can effectively inhibit goal attainment (Elliot & Church, 2003).

Self-handicapping could influence an ecotoxicologist's progress and quality of judgments in their research. My impression is that as the science of ecotoxicology unfolds, many talented and well-intended scientists remain averse to definitive experimentation that tests important paradigms and theories. Instead, careers are built upon many small and safe studies of the general form, "The [fill-in effect or change to be studied] of [fill-in the contaminant of interest] on [fill-in the test species] under [fill-in the test conditions]." While a certain number of such studies is invaluable to the field, many go sparsely cited or not cited at all. Borrowing the words of Platt (1964) in describing this behavior in a science, "We speak piously of taking measurements and making small studies that will 'add another brick to the temple of science.' Most such bricks just lie around the brickyard."

This habit of avoiding definitive tests of important concepts in ecotoxicology may be a product of self-handicapping. Innovation seems to be shunned by those most capable of breaking from the another-brick-in-the-wall behavior that carries little risk to self-esteem. One can do small another-brick-in-the-wall studies while claiming that ecosystems are too complex to understand adequately by any other approach or that regulators need these types of studies. The question becomes who then should risk their professional self-esteem to ask innovative questions or conduct definitive tests of central concepts in this science?

2.2.2.20   Imitation Including Informational Mimicry

Individuals of solitary species with minimal social learning must rely solely on their genetics and trail-and-error learning. Early humans, being a social species with extraordinary talents for trial-and-error learning, had the additional advantage over solitary species of being adept mimics of others who display adaptive behaviors (Rendell et al., 2010). "Evolution also favors a psychology that makes people more prone to imitate prestigious individuals who are like themselves even though this habit can result in maladaptive fads" (Richerson & Boyd, 2005). Humans use social learning selectively but sometimes their selectivity is imperfect (Kendal, Giraldeau, & Laland, 2009). The risk of relying too much on mimicry emerges when the individual misjudges who to imitate and under what conditions imitation is superior to individual trial-and-error learning (Giraldeau, Valone, & Templeton, 2002). Perhaps, the everyday form of information mimicry called gossip best illustrates this point. Most adults realize that a decision based on gossip is error prone unless augmented by trial-and-error testing of its soundness.

Subconscious rules balance information mimicry and trial-and-error testing by individuals making decisions (Richerson & Boyd, 2005). Reliance on mimicry is unwise in highly variable environments because of the high risk of using inappropriate or out-of-date information (Kendal et al., 2009; Rendell et al., 2010). At the other extreme, strict reliance on individual learning by fact gathering and evaluation can be inefficient and subject to segregation error. Mixed strategies applying both kinds of learning are the norm as discussed in Section 2.3.2.

Bird foraging studies suggested to Giraldeau et al. (2002) that trial-and-error learning alone makes sense only when drawing on both types of learning becomes impossible. Otherwise, mimicry is a valuable tool not to be neglected. But, suboptimal information cascades can result when an individual relies too heavily on social learning. Copying other individuals is likely the wrong strategy if the conveyed information possesses high uncertainty, that is, is easily misconstrued. An information cascade can precipitate decisions by individuals in groups who forego reliance on individual acquired information.[5] Given high uncertainty, information cascades can amplify suboptimal decisions. Given the *status quo* bias described earlier, an informational cascade that establishes a suboptimal decision or insubstantial paradigm in ecotoxicology could very likely result in a prolonged and wasteful exorcism later when conflicting evidence eventually comes to light.

In 1969, Sprague advocated using 96 hours as the exposure duration of choice in acute toxicity tests. "For 211 of 375 toxicity tests reviewed, acute lethal action apparently ceased with 4 days, although this tabulation may be biased..." He conscientiously stated the shortcomings of the associated data, acknowledging "room for healthy disagreement." His position was quickly adopted because no better suggestion was apparent, a standard was needed at that time, and 4 days fit conveniently into a work week. Newman and Clements (2008) reexamined this judgment 40 years later, pointing out

---

5. The informational cascade concept was introduced by economists to describe fads, fashions, and other changes associated with information movement through groups. Bikhchandani, Hirschleifer, and Welch (1992) state that "an informational cascade occurs when it is optimal for an individual, having observed the actions of those ahead of him, to follow the behavior of the preceding individual without regard to his own information." See their paper for more details including factors contributing to social conformity and informational cascades.

several issues. Framing the interpretation differently, the same results lead to the conclusion that the 96-hour duration was insufficient in more than 4 out of 10 tests. Also, many of the tests tabulated by Sprague (1969) were static toxicity tests, so the toxicant concentrations in the tanks likely decreased substantially during the 4 days of exposure. Despite much trial-and error evidence contradicting the 96-hour convention, how acute toxicity is examined is still influenced by the informational cascade catalyzed by Sprague's then-timely and thoughtful paper.

2.2.2.21   Leveling and Sharpening

This tendency emerges from instinctive rules of storytelling (Gilovich, 1991). A certain amount of leveling and heightening of information will always be present due to our predispositions designed to hold a listener's attention until they have heard our entire message. What is perceived as attention holding will be sharpened in our presentation and details subjectively judged as less interesting will be de-emphasized.

Although highly stylized, scientific communication is still storytelling. Leveling and sharpening of details can compromise cognition if the rules of scientific objectivity become secondary to those of good storytelling. Nonetheless, leveling and sharpening foster good scientific communication if applied with balance. Drawing an example from physics, both Einstein and Poincaré separately proposed versions of relativity theory but "In the competition for public recognition, the clarity and simplicity of Einstein's argument gave him an overwhelming advantage" (Dyson, 2006).

2.2.2.22   Probability Blindness

Piattelli-Palmarini (1994) sums up this next foible quite neatly, "... we are instinctively very poor evaluators of probability and equally poor at choosing between alternative possibilities..." He provides a simple rule governing estimated probabilities: "Any probabilistic intuition by anyone not specifically tutored in probability calculus has a greater than 50 percent chance of being wrong." This claim is supported by the large number of people who play the lottery despite the reality illustrated in Section 2.2.2.10. It also manifests in the opening example of estimating the probability of a tap water sample containing a pesticide. Another is the previous example of the maxim of relevance linguistic rule that muddles probabilistic reasoning enough to produce the conjunction fallacy.

Although acknowledging the difficulties arising from probability blindness, Gigerenzer tries to explain its roots and suggests a way of lessening its impact. He reasons that our ancestors dealt with counts and natural frequencies of things in their environment, not fractions and probabilities. (Natural frequencies are numbers such as those from enumerations during natural sampling. A natural frequency and its corresponding probability would be three people dying in a village of 1000 inhabitants and 0.003, respectively.) It is unsurprising that our native abilities are lacking when confronted with probabilities, a creation of the 17th century mathematicians, Baise Pascal and Pierre de Fermat (Gigerenzer, 2000). Gigerenzer explains that the natural interpretation of probability is correctness or degree of belief in single events. A probability envisioned as a quality emerging in the long run or with large numbers of samples from a population is a very specialized and recent context (Gigerenzer, 2000). As an example of probability serving as a measure of belief, a person might state that the probability of snow tomorrow is very low. It is much more difficult for an untrained or inattentive individual to understand what is meant by the probability of some particular outcome (coin toss) approaching 0.5 as the number of samples from a population becomes very large. As shown in Section 2.3.2, Gigerenzer's explanation of natural frequencies can be the key to reducing probability blindness in some instances.

Many ecotoxicologists and ecological risk assessors making judgments of plausibility or probability apply what is formally called global introspection. This approach to reaching an expert opinion involves drawing together all pertinent evidence, weighing that evidence based on perceived importance in hopes of identifying candidate causal links, and then making an expert decision about the probability of a particular cause for an effect (Lane, Kramer, Hutchinson, Jones, & Naranjo, 1987). Other's descriptions of global introspection lack the neutral tone attempted here. For example, Kramer's definition (Kramer, 1986) reflects the unreliability of this approach when applied in medical diagnostics. "[The assessor] makes a mental list of [factors that could possibly affect the causal link], weighs them according to some sense of their relative importance, mixes the complicated brew in a fashion somewhat reminiscent of the witches in *Macbeth*, and then spews forth a decision about the probability of ... causation."). As evidenced by the above materials, global introspection done without an awareness of probability blindness can produce very misleading decisions and inaccurate risk assessments. This point is strengthened by the discussion to follow about base rate neglect.

2.2.2.23 Base Rate Neglect (or Fallacy)

Base rate neglect was at the core of this chapter's opening example in which experts gave very poor estimates of the probability that a tap water sample contained pesticide. Kahneman and Tversky (1996) explain how this bias emerges in intuitive judgments about probabilities. They proposed the representativeness heuristic in which human judgments are based on "similarity or representativeness between evidence and possible outcomes" (Kahneman & Tversky, 1996), or in terms of probabilities, "when A is highly representative of B, the probability that A originates from B is judged to be high" (Tversky & Kahneman, 1974). Representativeness is weighed by how closely the correspondence is and that correspondence then becomes key to recognition. Event frequencies are much less important in this context. As a consequence, base frequencies are often neglected even when they are available for consideration.

In addition to base rate neglect in the introductory example, another bias, the isolation effect, was also present. Kahneman and Tversky (1979) proposed this propensity to disregard common components in different alternatives. The 1 in 1000 prior frequency of pesticide in a tap water sample was the common feature to both the positive and negative testing components of the question. Likely, those asked to provide an answer noted only that the tested sample came back from the laboratory as positive for the pesticide based on a test that had a high 95% chance of detecting the pesticide if present. From this line of reasoning and neglecting the common component (pesticide presence base rate), most judged that it was very likely that the pesticide was truly present. However, the following is the correct answer that takes into account the base rate and the two test error rates:

$$P(\text{Pesticide}|\text{Test}^+) = \frac{P(\text{Pesticide})P(\text{Test}^+|\text{Pesticide})}{P(\text{Pesticide})P(\text{Test}^+) + P(\sim\text{Pesticide})P(\text{Test}^+|\sim\text{Pesticide})}$$

$$P(\text{Pesticide}|\text{Test}^+) = \frac{(0.001)(0.95)}{(0.001)(0.95) + (0.999)(0.02)} = 0.0454$$

where $P(\text{Pesticide}|\text{Test}^+)$ = probability of the pesticide being present given a positive test,

> $P(\text{Pesticide})$ = the probability of the pesticide being present in a tap water sample,
> $P(\sim\text{Pesticide})$ = the probability of the pesticide not being in a tap water sample,
> $P(\text{Test}^+|\text{Pesticide})$ = the probability of a positive test given that the pesticide is present,
> $P(\text{Test}^+|\sim\text{Pesticide})$ = the probability of a positive test if the pesticide is not present.

The probability of the pesticide being truly present is 0.0454. Admittedly, these calculations require some knowledge of probability calculus before they can be applied to answering the question. Nonetheless, the application of global introspection under the influence of base rate neglect gives a very misleading conclusion. In Section 2.3.2.11, the use of natural frequencies will be illustrated for making these estimations easier.

2.2.2.24   Conjunction Fallacy

Tversky and Kahneman (1983) first framed this cognitive error in terms of formal probability conjunction. The probability of several things being true

(A and B) cannot be greater than the sum of the separate probabilities of each of them being true, $P(\text{A and B}) \leq P(\text{A}) + P(\text{B})$. It also follows that P (A and B) cannot be greater than $P(\text{A})$. For example, the combined probability of all three of species A, B, and C being harmed by a contaminant ($P(\text{A and B and C harmed})$) cannot exceed the probability of two of them being harmed ($P(\text{A and B harmed})$). Yet everyday heuristics lead humans astray of this rule, resulting in the conjunction fallacy. Continuing with our three at-risk species example, one might present an ecotoxicologist with the following scenario. Species A, B, and C have similar trophic ecologies and co-occur in a contaminated pond. A study finds that A and B are harmed by a contaminant in the pond. The ecotoxicologist is then asked to choose which is most probable for the contaminated pond that Species A and B are harmed or Species A, B, and C are all harmed. Most people will incorrectly pick the second choice but the first is more likely based on probability conjunction rules.

2.2.2.25 Certainty Effect

Kahneman and Tversky (1979) identified the certainty effect in which "people overweigh outcomes that are considered certain, relative to outcomes which are merely probable." They illustrate this bias with the preferences of responders for one of two options, one being a certainty. Choice 1 was a certain win of $2400. Choice 2 had three possible outcomes: a probability of 0.66 of winning $2500, a probability of 0.33 of winning $2400, and a probability of 0.01 of winning nothing. Eighty-two responders picked Choice 1 and only 18 picked Choice 2 despite the utility computation showing that Choice 2 had the highest expected value. The certainty of Choice 1 made it most desirable. Equally suboptimal decision-making occurs if losses are considered instead of gains. More weight is given to certainty in outcomes than less certain outcomes.

It is not hard to make a guess whether the certainty effect could color judgments by ecotoxicologists and ecological risk assessors. The bias is inherent in human decision-making, and in the absence of any effort to counteract its effect, it surely must influence their professional decisions.

2.2.2.26 Insensitivity to Sample Size

The representativeness heuristic, described above in discussions of base rate neglect, also influences how individuals include sample size in their judgments and decisions. Just as representativeness was weighed more heavily than base rate frequencies, sample size also is unconsciously deemed less important than representativeness in judgments. Tversky and Kahneman (1974) give the example of probabilities of an average height of men being greater than 6 feet tall if samples of sizes 1000, 100, or 10 men were taken.

Neglecting the impact of sample size, responders estimated the same probabilities from the three different sample sizes.

Given the traditional frequentist statistics training in the field, this issue is likely less prevalent in ecotoxicology than in the general population studied by psychologists. However, it is certainly still present and requires attention during interpretation of data and assessing risk. Activities in which it likely emerges unnoticed are statistical estimation and testing. Sample size is rarely addressed sufficiently in most ecotoxicology studies despite the importance of knowing how many samples are required to obtain adequate estimation precision or statistical power during testing. Insensitivity to sample size seems to be present, although in a more subtle way, in ecotoxicology and risk assessment. Estimation precision or power analyses done prior to study would lead to a fuller appreciation of sample size impact.

2.2.2.27 Visual Misperception

This innate human foible is not a formal heuristic *per se*, but it can sway decisions and judgments. It is discussed here because of the central role played by visual tools (graphs, plots, and images) in interpreting and communicating evidence.

Visual misperception is often demonstrated with the St. Louis Gateway Arch illusion (Fig. 2.3, top panel). The arch designer took advantage of how our brains process visual evidence when gaging dimensions of objects placed on a horizon. The arch tapers in a very specific way from ground to apex to give the illusion of stretching upward for a distance greater than the horizontal distance between its two legs at its base (Osserman, 2010). Actually, the vertical distance from ground to peak and horizontal distance from leg to leg are the same.

How visual presentations of technical evidence are interpreted is also subject to misperceptions. The bottom panel of Fig. 2.3 illustrates this point. The two curves appear to be converging on the right side of the figure; yet, the difference in *y*-values along the two curves (double arrowhead vertical lines) is the same at all points. The problem emerges because one's eye gages closeness by the shortest distance between the lines, not the vertical distance between them.

Other difficulties arise such as our general difficulty in judging evidence if scales on plots are not arithmetic. (Given the subject of this book, it is important to emphasize before continuing that logarithmic scales are often better than arithmetic scales when communicating odds such as those of an adverse effect or those associated with updating with Bayes factors (D'Agostini, 2010)). Confusion arising from nonarithmetic scales can be illustrated with the fit of saltwater acute sensitivities to tributyltin (TBT) to a log-normal model by Hall, Scott, Killen, and Unger (2000). The general intent of the paper was estimation of environmental concentrations of TBT

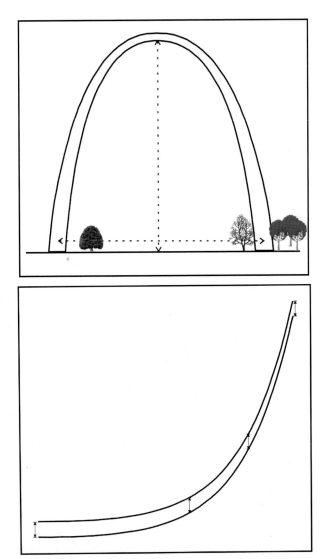

FIGURE 2.3 Misperceptions resulting from how our minds process visual evidence. The St. Louis Arch is designed intentionally to give the false visual impression that it is taller than it is wide (top panel). Visual misperceptions can also confuse judgments from plots of evidence (bottom panel, modified from Kolata (1984)).

that allowed "protection of 90% of the species 90% of the time (10th percentile)" (Hall et al., 2000) using methods sanctioned by the Society of Environmental Toxicology and Chemistry Aquatic Risk Assessment and Mitigation Dialogue Group. The top panel of Fig. 2.4 depicts their figure 4

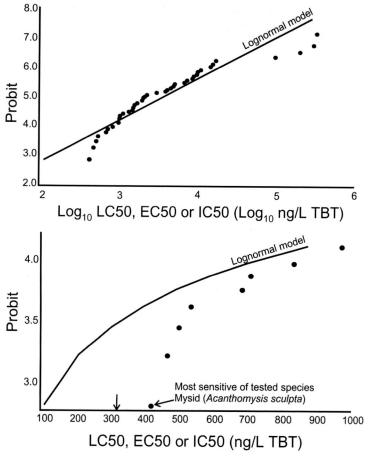

FIGURE 2.4 The influence of axis scaling on misinterpretation of acute TBT effect evidence. The *top panel* displays the evidence as depicted by Hall et al. (2000) with concentration converted to logarithms. The data conformity to the log-normal model (solid line) was judged adequate by the authors and a concentration thought to protect 90% of species 90% of the time was estimated to be 320 ng/L. The *bottom panel* uses arithmetic units for concentrations, showing clearly the inadequate fit of the model to the data in the relevant low range of concentrations. Incongruously, all 43 LC50 values were higher than the concentration predicted to protect all but 10% of exposed species (vertical *arrow*).

which uses a log-scaled *x*-axis and a probability scaled *y*-axis. The regression fit to an assumed log-normal model is shown as a solid line. Although the lower left side of the plot was of most interest, the authors used the entire range of observations to estimate a notionally protective 10th percentile of 320 ng/L (depicted with an arrow on the *x*-axis of the bottom panel). Strong influence of a *status quo* bias was assumed for the authors because they overlooked the nonrandom pattern of the observations around the log-normal model predictions. The visual appearance with logarithmic scaling contributed substantially to this oversight: the entire set of log-transformed data was visually suggestive of a roughly straight line. However, the bottom panel of Fig. 2.4 shows the relevant portion of the data set using an arithmetic scale for concentration. The inadequacy of the log-normal model for its intended purpose becomes immediately obvious from this replotting. Wheeler, Grist, Leung, and Crane (2002) expressed similar concern about another paper by these authors (Hall, Scott, & Killen, 1998) and found that a logistic model provided a better fit than a lognormal one for the data.

2.3 CONCLUSION

2.3.1 The Wisdom of Cognitive Insecurity

What is the overall message of this chapter? Our understanding of the natural science of pollution and our abilities to determine contaminant levels and impacts have grown impressively during the last four decades. We should allow ourselves a feeling of satisfaction with our collectively well-done job in this regard. Unfortunately, our understanding of the impact of psychological foibles on our scientific and assessment judgments has failed to grow at the same rate despite its importance and the urgency for improving inferences.

The argument could be forwarded that ecotoxicologists and ecological risk assessors are trained professionals who are less subject than most to the stumbling blocks just described. But examples provided throughout this chapter do not support this argument. The water quality survey described at the beginning of this chapter disclosed that trained environmental scientists and students are susceptible to base rate neglect, one very telling manifestation of probability blindness. Availability as influenced by social accentuation was illustrated with the saga of Hayes's challenging of the environmental safety of atrazine. Excessive value centrality bias was ascribed to Ronald Fisher, the originator of statistical methods currently used by most ecotoxicologists and risk assessors. Errors due to visual misperception were illustrated with species sensitivity distribution analyses.

Still other studies support the broader proposition that scientists are not immune to commonplace errors in judgment. Mahoney and DeMonbreun (1977) tested PhD psychologists, PhD physical scientists, and conservative Protestant ministers relative to their abilities to avoid hypothesis confirmation bias. Using the already-described methods of Wason (1960), they found no overall significant differences among groups' abilities to avoid confirmation bias. Scientists were no different from ministers in applying the hypothesis falsification approach. Indeed, more psychologists (93%) and physical scientist (93%) reconfirmed previously falsified hypotheses than did ministers (53%). When told his hypothesis was wrong, one psychologist even countered that "my rule is as good as your rule." This broad study of scientists' susceptibility to logical error is consistent with that of Barber (1961) who demonstrated a general resistance of scientists to new concepts and discoveries.

There is no evidence from these studies that scientists, including ecotoxicologists, are better equipped to avoid cognitive biases than nonscientists. One could counter that despite this being the case, the peer review process of manuscripts and grant proposals will catch many such errors. However, Horton (2003) observes that "Peer review tells us about the acceptability, not credibility, of a new finding." Peer review often acts more as a tool to separate the acceptability of a finding relative to existing theory or facts than as a tool for discerning the evidence-based credibility of a new finding (Barber, 1961). Similarly, Dyson (2006) observes, "If a committee of scientific experts selects research projects by majority vote, projects in fashionable fields are supported while those in unfashionable fields are not."

Given the conditions just discussed, it would seem wise to explore more thoughtfully the influence of these biases and foibles in ecotoxicology. Although understandable during the early stages of the field, continued inattention to these influences would constitute willful neglect and from the context conveyed in Chapter 1, The Emerging Importance of Pollution, a dereliction of the field's broad responsibilities to society. The present stage of scientific evolution of ecotoxicology and the critical needs of the 21st century make a strong case for the movement of the field out of its present nonage to a more mature and responsible stage. Some insights and methods from clinical, business, and social sciences for lessening the impact of the cognitive biases are briefly discussed below in the hope of stimulating more interest from ecotoxicologists and ecological risk assessors.

2.3.2   Suggestions for Minimizing Cognitive Biases

The overarching approach to lessening biases is to cultivate an active awareness of their existence. Learning about biases and then applying this understanding to moderate judgments would reduce the number and severity of mistakes made in ecotoxicology and ecological risk assessment. Beyond this recommendation, some specific points are made below for some of the biases and foibles already discussed. Materials in other chapters also provide useful insights.

2.3.2.1   Overconfidence Bias

Overconfidence bias emerges in many situations, and it follows that the best ways of reducing its influence will depend on the specifics of the situation. An individual making a judgment might be subject to naïve realism, that being, believing that their perceptions and judgments are objective, and that any *sensible* person will share these views (Liberman, Minson, Bryan, & Ross, 2012) According to Liberman et al. (2012), naïve realism and the related underweighting or ignoring of other's views/opinions results from availability and anchoring. An individual's reasons and evidence behind their judgment are more readily available for recall than those of peers. Once an individual makes a decision, it and the associated evidence then become the anchoring point from which the decisions and views of others are weighed. One consequence of naïve realism is that an individual making a decision generally gives more weight to their own views than to those of others. Underweighting of judgments by peers will become more serious the more their judgments diverge from one's own (Liberman et al., 2012) or when a large difference in knowledge of the subject is believed to exist (Yanis, 2004). When views are not highly divergent and both the decision maker and their advisor are believed to have equal abilities and knowledge of a subject, Yanis (2004) gives a general weighting of 70% to one's own opinion and only 30% to that of the advisor. This propensity to underweight can even reach the extreme of completely ignoring very different judgments/views of dissenters. The errors resulting from and ways of reducing the influence of naive realism by effective peer advice will be detailed in Section 6.2.

Even more attention is required when combining judgments of several individuals as in the cases of expert elicitation or peer review of a manuscript. Simply warning decision makers about the dangers of overconfidence is ineffective according to Gigerenzer (1991). If judgments involve quantitative estimation such as the probability of something being true, Gigerenzer (1991) suggests that using natural frequencies in place of probabilities (i.e., 1 in 100 instead of 0.01) can reduce overconfidence bias; however, Kahneman and Tversky (1996) discount the value of Gigerenzer's suggestion. A formal training or calibration exercise might be warranted in some cases (Fischhoff et al., 1977). In a formal calibration exercise, decision makers or experts are given a series of judgments to make during a preliminary testing in which the correct judgment/decision is known to the tester. The expert's judgment and associated confidence in each decision are noted. A calibration curve of responses versus proportion correct is generated to calibrate each individual expert. The resulting estimates of each individual expert's overconfidence are then used to weigh their judgments during the combining of all experts' judgments.

2.3.2.2   *Availability (Ease of Recall) as Influenced by Vividness or Emotional Content, Confirmation Bias, Anchoring, Endowment Effect, and Status Quo Bias*

In my opinion, hypotheses championed by individual scientists are nearly always imbued with a certain amount of emotional content. Put bluntly, tenure and grant monies are awarded based on how often a scientist is successful, i.e., correct in their research hypotheses. Despite the reality of such a flawed award system, emotional attachment impedes progress in objectively judging the relative strengths of hypotheses and providing sound advice to the regulatory community. As a consequence, confirmation bias, anchoring, and resistance to rigorous testing of a favored hypothesis manifest to varying degrees. For science conducted in a healthy way, no better advice for avoiding these impediments has emerged since Chamberlin (1897) suggested the following approach:

> *[The method of multiple working hypotheses] differs from the simple working hypothesis in that it distributes the effort and divides the affections ... In developing the multiple hypotheses, the effort is to bring up into view every rational explanation of the phenomenon in hand and to develop every tenable hypothesis relative to its nature, cause or origin, and to give to all of these as impartially as possible a working form and a due place in the investigation. The investigator thus becomes the parent of a family of hypotheses; and by his parental relations to all is morally forbidden to fasten his affections unduly upon any one. In the very nature of the case, the chief danger that springs from affection is counteracted. Where some of the hypotheses have been already proposed and used, while others are the investigator's own creation, a natural difficulty arises, but the right use of the method requires the impartial adoption of all alike into the working family. The investigator thus at the outset puts himself in cordial sympathy and in parental relations (of adoption, if not of authorship,) with every hypothesis that is at all applicable to the case under investigation.*

Instead of becoming professionally vested in a hypothesis, a researcher becomes vested and skillful in this superior way of judging hypotheses. Subjective attachments to explanations and hypotheses are reduced in this manner. This approach appears to also be as valuable for ecological risk assessment activities as for scientific ones.

2.3.2.3   Availability (Ease of Recall) as Influenced by Social Accentuation

The atrazine example provided earlier suggests a partial remedy for this bias. The recent controversy surrounding the harmfulness of atrazine fascinated the ecotoxicology community but its attendant social accentuation distracted the community from the serious business of objectively evaluating the true danger associated with this herbicide. Such distractions should be moderated, so that much more objective evidence such as that of Rohr and McCoy (2010) is given fuller consideration. Personally, I intentionally avoided attending the many combative talks associated with this controversy, and instead, focused on the peer-reviewed publications as they appeared. It is unfortunate that the professional societies that hosted these talks did not act more assertively to reduce the social accentuation surrounding the issue.

2.3.2.4 Framing Error

How a scientific or risk hypothesis is framed influences judgment and more effort is needed to frame hypotheses as neutrally as possible. Generally, the method of multiple working hypotheses (see also Section 3.2.2) reduces the influence of framing during research efforts. Because ecotoxicology has legitimate technological, in addition to scientific goals, there is a risk that putatively scientific endeavors will misframe experiments in the scientifically inappropriate context of validation instead of falsification or abduction. Validation is very appropriate for assessing the soundness of a technology for regulatory purposes but is inappropriate for framing and testing neutral scientific questions.

Thoughtful framing can facilitate practical judgments by individuals too. Johnson and Goldstein (2003) provide a good illustration of how framing allowed individuals to easily select options in accordance with their beliefs. They noted that most Americans (85%) favor organ donation but only a minority (28%) signed donor cards. By comparing European donor programs, they discovered that the way that the organ donor option was presented allowed individuals in some countries to effortlessly choose in accordance with their true desires. In a country in which organ donation was the default choice, approximately 86%−100% chose to donate. Only 4%−28% chose to donate if the individual had to actively check off on a driver's license application that they were willing to donate. Active selection involved momentary unpleasant thoughts about one's eventual death. For that moment, attention shifts away from a genuine desire to help others to an uncomfortable state of emotional aversion. Proper framing as an opt-out instead of an opt-in default choice eliminated this impediment. An effective framing example pertinent to environmental decision-making was already discussed relative to climate change. A charge given to individual or groups of decision makers might use the term, "climate change" instead of "global warming" to reduce bias in final judgments.

2.3.2.5 Acquiescence

Piattelli-Palmarini (1994) defined this bias as the acceptance of a solution to a problem or explanation of an observation based on the way it was originally presented. As is the case for many of the biases and foibles described here, the method of multiple working hypotheses or abduction detailed in Chapter 3, Human Reasoning: Within Scientific Traditions and Rules, can reduce the danger of falling victim to acquiescence error.

2.3.2.6 Risk Aversion

A certain level of anxiety about the correctness of a future judgment can encourage careful evidence gathering before making a decision. An impediment emerges only if excessive anxiety creates risk aversion with its attendant *ad hoc* justifications and bolstering of suboptimal decisions. Therefore, it would be useful to understand the features of decision-making that contribute to appropriate or excessive amounts of anxiety. Janis and Mann (1977) highlight the following influences on predecisional vigilance.

- Appropriate levels of vigilance are likely if there are only minor obvious advantages of the preferred choice over others. Although fostering healthy levels of vigilance, this condition can foster a *status quo* bias if the preferred choice is that associated with the existing state of affairs. The danger of bolstering and neglect of alternatives emerges if one decision appears initially to be the only *reasonable* one. In such instances, special attention is needed to determine if the apparent acceptability of that preferred choice is due to the cursory manner in which the initial assessment of all choices was made.
- Healthy vigilance is also fostered if clear, negative consequences of the wrong decision will manifest soon after the decision is made. There is a tendency to be less vigilant during decision-making if there are no anticipated immediate consequences of a suboptimal decision. For example, a natural resource damage assessment might carefully gather evidence of the number of individual birds killed during an oil spill and then diligently decide on the legal action warranted by the estimated damage. However, decisions associated with ecosystem damage from that same oil spill might be made with less diligence because the consequences of a wrong decision might not be obvious for decades.
- Appropriate vigilance is fostered if key people, such as other scientists studying the same topic or stakeholders engaged in a risk assessment, envision a judgment as important and something to be adhered to. A decision becomes embedded in the social network in which the decision maker exists so much so that it would be difficult to modify the decision later if found to be suboptimal (However, the problems of satisficing, polythink and groupthink can emerge in a group of decision makers under these conditions as will be discussed in Sections 6.3 and 6.4). At the other extreme, decision-making is compromised if a decision can be reversed quickly without any consequences to the decision maker or if the decision appears not to have the attention of the decision maker's social network. Low risk from a bad decision results in a low levels of vigilance when selecting among choices.
- Vigilance rises to the appropriate level for sound, well-reasoned choices if people important to the decision maker express patience and anticipation of a thorough and comprehensive evaluation process. Otherwise, perceived pressure from significant individuals within a decision makers' social network can encourage premature and potentially suboptimal decisions. For example, a mandate at the end of a protracted and contentious legal action between a responsible party and regulators could force

ecological risk assessors into a situation in which rushed suboptimal decisions are likely to occur.

- Undue anxiety during decision-making can be reduced if the decision maker anticipates that new evidence could emerge during evidence sorting that will make the soundness or unsoundness of a favored decision clearer. The perception that no additional information is likely to emerge that will influence a decision can lessen the desire to seek more evidence. Bolstering can then emerge to end the evidence gathering process. A problem can also emerge if information is brokered, that is, if information is trivialized or excluded from consideration by those purportedly supporting the decision maker.

Any action that encourages balanced vigilance while avoiding excessive predecision anxiety will increase the quality of judgments by ecotoxicologists and ecological risk assessors. In many cases, attention to the five factors bulleted above can improve decision-making. Periodic consultation with a valued peer or mentor can help individual researchers (e.g., open and frank discussions of preliminary findings at national meetings or "friendly fire" manuscript reviews prior to formal submission).

Other tools might help an individual trying to make a decision within a social group or network. A moderator aware of the above points might guide decision-making to avoid conditions fostering bad decisions. For example, a moderator can reduce bolstering by framing a decision as open ended and periodically reminding the group members of this condition. A directive agent giving the charge to an individual, network, or committee might also avoid eliminating or restricting freedom of choice. Such loss of freedom might be further reduced in committees by assigning a devil's advocate whose stated role is to challenge the experts during deliberations. Although sometimes difficult to tolerate, an individual ecotoxicologist might use a permutation of this approach by asking a professional opponent to review a manuscript before submitting it for formal review. To avoid owning a hypothesis and keeping the explanation "decision" open, the final publication by this same ecotoxicologist might include discussion of alternative explanations for the results and experiments that might rigorously test the proposed explanation. This also reduces any emotional bias associated with one-sided events by transforming the event into a two-sided one.

2.3.2.7 Self-Handicapping

Gilovich (1991) identifies two kinds of self-handicapping: real and feigned. To avoid negative judgments by others of one's performance, an individual might subconsciously create obstacles to achieving or completing a task. This is real self-handicapping. Feigned handicapping involves duplicitous claims about the difficulties in completing a task with the goal of controlling expectations of others in case of suboptimal performance. Real handicapping seems the most relevant form to address here, and it will be explored relative to the important issue of the characteristic failure to explore more innovative theories, hypotheses, or approaches in ecotoxicology.

Suggesting that young scientists should be most innovative scientists, Max Plank famously quipped that a new idea is accepted only after old scientists die and young scientists comfortable with the new idea become opinion leaders. Along this line of thinking, Loeb (2010) recently urged young scientists to pay more attention to high-risk research despite the dangers to their standing. Contrary to this position, Wray (2003) suggests that the risk is too high for young scientists and middle-aged scientists are best positioned to conduct high-risk innovative science. This incongruity will be examined here because innovation appears to be critically needed in ecotoxicology and self-handicapping seems to contribute to the lack of innovation.

Self-handicapping influences decisions by ecotoxicologists of various ilks relative to challenging current paradigms and proposing new ones in their research. The qualities of those who should innovate in ecotoxicology are worth elucidating and might help identify individuals most likely practicing self-handicapping relative to innovation. Is innovation the role of young scientists, established scientists, or leaders as defined by professional society roles or position in a regulatory institution? Once the most plausible innovators are identified, attempts can begin to reduce self-handicapping by those individuals.

To imitate those around you instead of confronting them is an evolved human trait (Gigerenzer, 2008). Imitation of prestigious individuals or those like ourselves is anticipated behavior of a sensible person wishing to forego the risk of individual trial-and-error reasoning (Richerson & Boyd, 2005). Such innate traits influence intellectual performance and sometimes put at risk (i.e., stereotype threat) those daring to move outside a socially defined range of imitative behaviors (Steele, 1997). Focusing these behaviors specifically on scientific research activities, it is philosophically acceptable, yet socially disruptive, to break from a pattern of imitation. In practice, proposing a novel explanation or approach threatens the standing of those vested in an established one. An innovator in any science risks their standing in their field and career viability at their home institutions if resistance by vested individuals is strong enough to bring into question the innovator's credentials or credibility. This theme is repeated throughout the philosophy of science literature.

Certain men or classes of men come to be the accepted guardians and transmitters—instructors—of established doctrines. To question the beliefs is to question their authority; to accept the beliefs is evidence of loyalty to the powers that be, a proof of good citizenship. Passivity, docility, acquiescence, come to be primal intellectual virtues. Facts and events presenting novelty and variety are slighted, or are sheared down till they fit into the Procrustean bed of habitual belief.

Dewey (1910)

"Almost always (always, in the mature sciences) the acceptance of a new theory demands the rejection of an older one. In the realm of theory, innovation is thus necessarily destructive as well as constructive."

Kuhn (1977)

It is hard to reconcile a pattern of active resistance to scientific change with Planck's curmudgeonly comment that scientific innovation is most facilitated by the young. The average young scientist or regulator has the least amount of accumulated credentials and position security to risk in challenging established concepts and methods. One could counter this opinion with anecdotes. As an oft-cited instance, Stanley Miller's classic work on the origins of life and primordial atmospheric chemistry was the product of a young scientist. However, that young scientist was working in the laboratory of Nobel laureate, Harold Urey, whose scientific clout provided ample protection for him. Next, one could propose that the most secure and skilled scientists should take on the role of innovators; however, established individuals are often highly vested in current concepts and methods. Any suggestion that opinion leaders carry the responsibility is also untenable. Opinion leaders are rarely good innovators for good reasons. They are selected as leaders in a science because they best reflect traditional norms. "In systems with traditional norms, the opinion leaders are usually a separate set of individuals from the innovators" (Rogers, 1995). The next common argument is that only rare, gifted individuals are responsible for innovation; however, this seems an obvious attempt at self-handicapping by those not wishing to risk their own professional status by attempting to innovate.

My conclusion from the above discussion is that any ecotoxicologist with the characteristics about to be described has an obligation to conduct innovative science. An innovative scientist is usually someone present during a Kuhnian crisis (paradigm shift) who becomes aware of the need for change.[6] That individual needs to be trained in, and open to, innovative reasoning as opposed to imitative or rote reasoning. They tend to be individuals who explore outside of, and draw ideas from, other disciplines or groups.[7]

---

6. As discussed in more detail in Chapters 3, 5, and 7, Thomas Kuhn's *The Structure of Scientific Revolutions* (1970) examined the history of scientific paradigms, finding that each paradigm went through a predictable cycle of displacing an earlier failing paradigm, accumulation of facts around the new paradigm, refinement of the new paradigm, eventual appearance of anomalous facts inconsistent with the paradigm, and increased questioning of the paradigm until it is replaced by a newer paradigm more consistent with current evidence. The event during which a paradigm begins to be questioned enough to prompt its eventual replacement is called a (Kuhnian) paradigm shift or crisis. See also Conant and Haugeland (2000).

7. These individuals will be discussed more formally in Chapters 6–8, as individuals who are open to heterophilic exchange. The premise that solutions to problems emerging within a group (e.g., the community of ecotoxicologists) often can be found in sources outside of that group by individuals open to heterophilic exchange is called the strength of weak ties theory.

One example is Robert May, who adapted mathematical techniques learned during his physics training to do ground breaking work on population dynamics (May, 1974). Another is Lothar Wegener, a meteorologist by training, who introduced continental drift theory into the field of geology, which was firmly entrenched in the contractionism explanation of the Earth's surface features. A recent team of this kind of innovator in ecotoxicology includes Donald McKay and colleagues whose fugacity modeling, mountain cold trapping, global distillation and fractionation, and aquivalence innovations draw freely from the diverse disciplines of thermodynamics, chemical engineering, and marine sciences. As an additional characteristic, a successful innovator is often secure enough in their scientific discipline to accept the associated risk, although as already discussed, the perception of security and risk can be subjective. Finally, an innovator is seldom an opinion leader because opinion leaders are generally chosen based on conformity to the established beliefs of a discipline: opinion leaders succeed by being in-group members.

2.3.2.8   Imitation Including Informational Mimicry

In his famous 1637 *Discourse on Method*, Descartes set down as his first personal rule of scientific reasoning "... never to accept anything as true that I did not know to be evidently so; that is to say, carefully avoid precipitancy and prejudice." This still seems a wise rule if practiced intelligently. However, in instances in which individual trial-by-error judging is prone to unacceptably high levels of error, imitation is a necessary compromise (Gigerenzer, 2008; Kendal et al., 2009). Imitation is reasonable in a stable environment but risk prone in a changing one. The unstable environment in any science is a period of paradigm crisis. In such times, it might be best to conscientiously adopt a trust-but-verify approach, withdrawing belief if a heretofore accepted paradigm or method proves inconsistent with mounting evidence. According to Gigerenzer (2008), imitation is also dangerous if "the environment is noisy and consequences are not immediate, that is, it is hard or time-consuming to figure out whether a choice is good or bad...". Translating this to the context of a science, imitation is unwise if it is ambiguous that a chosen judgment or explanation is the best of all plausible alternatives.

2.3.2.9   Leveling and Sharpening: Availability (Influenced by Social Attenuation or Familiarity)

Gilovich (1991) provides sound advice that seems pertinent to ecotoxicology, "Be skeptical in direct proportion to the remoteness of the message." The 2014 *The New Yorker* article about the personal struggle of Hayes against the powerful agrochemical industry involved artful leveling of some facts combined with sharpening of others via social attenuation.[8] This required more skepticism from the reader about conclusions than did the meta-analysis by Rohr and McCoy (2010). Other examples are easily found. Barker's, 1997 book, *And The Waters Turned to Blood*, takes the vantage of one researcher who ostensibly overcame formidable odds to identify the true cause of fish kills on the North Carolina coast. The back of Barker's book proclaims, "In the coastal waters of North Carolina—and now extending as far north as the Chesapeake Bay area—a mysterious and deadly aquatic organism named *Pfiesteria piscicida* (scientists call it 'the cell from hell') threatens to unleash an environmental nightmare and human tragedy of catastrophic proportions ... At the very center of this narrative is the heroic effort of Dr. JoAnn Burkholder and her colleagues, embattled and dedicated scientists confronting medical, political, and corporate power to understand and conquer this new scourge before it claims more victims." A publication by Paerl, Pinckney, Fear, and Peierls (1998) hypothesizing another explanation for the fish kills elicited the blunt comment from the book's central character, "[Pearl et al.] made serious errors of omission, germane from the perspective of science ethics..." (Burkholder, Mallin, & Glasgow, 1999). The anthropologist, David Grifith (1999) suggested that the risk posed by *Pfiesteria* was exaggerated, prompting counter claims by Burkholder and Glasgow (1999) of "falsely based personal attacks." Social accentuation inserted substantial bias into the scientific process. Leveling and sharpening in the secondary literature and popular press jumbled attempts by coastal resource managers to resolve the issue. Later Bayesian analyses (Borsuk, Stow, & Reckhow, 2004; Newman, Zhao, & Carriger, 2007) indicated that near anoxic bottom waters was another plausible cause of the fish kills in these highly eutrophic coastal waters.

Time is also germane in the context of both storytelling tendencies and availability of evidence for recall. Conscientious ecotoxicologists keep abreast of recent publications and findings in their science; they are often familiar with their fellow ecotoxicologist who are producing these publications. As a result, evidence published in the last 7 or 10 years comes more easily to mind than that in older publications. Older publications are summarized in secondary sources which, as suggested by Gilovich (1991) above, are not as reliable as the original work due to leveling and sharpening. An ecotoxicologist might make a habit of reading back further in the primary literature than typical to reduce this influence of time on evidence weighing during decision-making, especially during a Kuhnian crisis.

---

8. To be fair, the substance of this article is also enhanced by fluency, "the subjective experience of ease or difficulty associated with a mental process" (Oppenheimer, 2008). The fluency heuristic states that, all else being equal, highest credibility or value is given to information that is communicated with the most fluency (Oppenheimer, 2008). The 2014 article in *The New Yorker* was well crafted and the conveyed information gained substantial credibility due to the writer's skills. The same can be said for Barker's book about to be described.

2.3.2.10   Probability Blindness

Some mechanisms for reducing probability blindness are available such as that described below for base rate neglect. However, diligence is necessary as our limited ability to intuitively work with probabilities often causes undetected errors. Proper decision-making involving probabilities requires statistical training. Such statistical training should go beyond rote frequentist estimation and significance testing to include probability and permutation theory, and a general understanding of information-theoretic and Bayesian methods.

2.3.2.11   Base Rate Neglect

Awareness of its existence can reduce the risk of base rate neglect. However, it is often difficult to move from awareness to the next step of properly including base rate. A Bayesian equation including base rates was illustrated in Section 2.2.2.23, and some statistically trained ecotoxicologists might quickly find the appropriate equation for doing computations for other situations. For those who prefer another solution, recasting evidence in terms of natural frequencies can make it easier to do the necessary estimations (Gigerenzer, 1991, 2000). Fig. 2.5 recasts the tap water contaminant example used to open this chapter in terms of natural frequencies. The same answer can be obtained with these more intuitive natural frequencies as that calculated with the Bayesian equation.

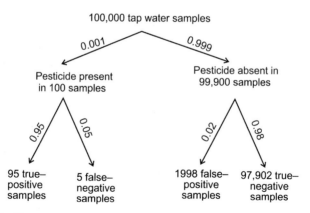

FIGURE 2.5   The opening question about pesticide presence in tap water can be answered using natural frequencies instead of probabilities. The example begins by envisioning 100,000 tap water samples. We are told that 1 in every 1000 will actually contain detectable amounts of the pesticide so 100 samples of the 100,000 will have the pesticide in them and the rest (99,900) will not. Of the 100 with the pesticide in them, 95 of them will test positive and 5 will not. Of the 99,900 that contain no pesticide, only 2%, or 1998, will produce a false-positive test result. So, there were 95 true and 1998 false-positive samples. That means 95/(95 + 1998) or 0.045 of the samples that tested positive will actually have the pesticide in them. This is the same answer obtained with the equation used earlier; however, the use of natural frequencies instead of probabilities makes the solution much easier to reach for someone lacking statistical expertise.

2.3.2.12   Visual Misperception

Presentation of evidence in diagrams and graphs can be misleading as illustrated by Fig. 2.3. There are publications that give guidance for visually presenting evidence to minimize such problems. Chapter 4 in Cleveland (1994) discusses graphical perception. Kolata (1984) is a brief guidance publication from which the bottom panel of Fig. 2.3 was derived. Cleveland and McGill (1985) and Wainer (1984) are two others that demonstrate the importance of picking the correct plot aspect, scales, and axes.

REFERENCES

Adams, J. (1995). *Risk*. London: UCL Press Limited.

Aquatic Dialogue Group. (1994). *Aquatic dialogue group: Pesticide risk assessment & mitigation*. Pensacola: SETAC Press.

Aviv, R. (2014). A valuable reputation. *After Tyrone Hayes said that a chemical was harmful, its maker pursued him. The New Yorker*, retrieved 10.02.14 Condé Nast.

Bacon, F. (1620). *Advancement of learning and novum organum*. New York: Wiley Book Co. (reprint 1944).

Balmford, A. (2013). Pollution, politics, and vultures. *Science, 339*, 653−654.

Barber, B. (1961). Resistance by scientists to scientific discovery. *Science, 134*, 596−602.

Barker, R. (1997). *And the waters turned to blood*. New York: Touchstone.

Beggan, J. K. (1992). On the social nature of nonsocial perception: the mere ownership effect. *Journal of Personality and Social Psychology, 62*, 229−237.

Bikhchandani, S., Hirschleifer, D., & Welch, I. (1992). A theory of fads, fashion, custom, and cultural change as informational cascades. *Journal of Political Economy, 100*, 992−1026.

Bojas, G. J., & Doran, K. B. (2013). *Prizes and productivity: How winning the fields medal affects scientific output. Working Paper #022*. University of Notre Dame, Department of Economics.

Borenstein, M., Hedges, L. V., Higgins, J. P. T., & Rothstein, H. R. (2009). *Introduction to meta-analysis*. Chichester: John Wiley & Sons, Ltd.

Borsuk, M. E., Stow, C. A., & Reckhow, K. H. (2004). A Bayesian network of eutrophication models for synthesis, prediction, and uncertainty analysis. *Ecological Modeling, 173*, 219−239.

Burkholder, J. M., & Glasgow, H. B., Jr. (1999). Science ethics and its role in early suppression of the *Pfiesteria* issue. *Human Organization, 58*, 443−455.

Burkholder, J. M., Mallin, M. A., & Glasgow, H. B., Jr. (1999). Fish kills, bottom-water hypoxia, and the toxic *Pfiesteria* complex in the Neuse Rover and estuary. *Marine Ecology Progress Series, 179*, 301−310.

Buss, D. M., Haselton, M. G., Shackelford, T. K., Bleske, A. L., & Wakefield, J. C. (1998). Adaptations, exaptations, and spandrels. *American Psychologist, 5*, 533−548.

Cairns, J., Jr. (1983). Are single species toxicity tests alone adequate for estimating environmental hazard? *Hydrobiologia, 100*, 47−57.

Cairns, J., Jr. (1986). What is meant by validation of predictions based on laboratory toxicity tests? *Hydrobiologia, 137*, 171−178.

Cairns, J., Jr. (1992). The threshold problem in ecotoxicology. *Ecotoxicology, 1*, 3−16.

Cairns, J., Jr., & Pratt, J. R. (1993). Trends in ecotoxicology. *The Science of the Total Environment, 134*(Supplement 1), 7−22.

Calow, P., Sibly, R. M., & Forbes, V. (1997). Risk assessment on the basis of simplified life-history scenarios. *Environmental Toxicology and Chemistry, 16*, 1983−1989.

Chamberlin, T. C. (1897). The method of multiple working hypotheses. *Journal of Geology, 5*, 837−848.

Cleveland, W. S. (1994). *The elements of graphing data.* Summit, NJ: Hobart Press.

Cleveland, W. S., & McGill, R. (1985). Graphical perception and graphical methods for analyzing scientific data. *Science, 229*, 828−833.

Conant, J., & Haugeland, J. (Eds.), (2000). *The road since structure. Thomas S. Kuhn.* Chicago: The University of Chicago Press.

Cox, D. R., & Oakes, D. (1984). *Analysis of survival data.* New York: Chapman & Hall.

Crane, M., Newman, M. C., Chapman, P. F., & Fenlon, J. (Eds.), (2001). *Risk assessment with time to event models.* Boca Raton: CRC/Lewis Publishers.

Croskerry, P. (2003). The importance of cognitive errors in diagnosis and strategies to minimize them. *Academic Medicine, 78*, 775−780.

Cyert, R. M., & March, J. G. (1963). *A behavioral theory of the firm.* Englewood Cliffs, NJ: Prentice-Hall.

D'Agostini, G. (2010). A defense of Colombo (and the use of Bayesian inference in forensics): A multilevel introduction to probabilistic reasoning. <http://arvix.org/abs/1003.2086v2>.

Dawson, N. V., & Arkes, H. R. (1987). Systematic errors in medical decision making: judgment limitations. *Journal of General Internal Medicine, 2*, 183−187.

De Martino, B., Kumaran, D., Seymour, B., & Dolan, R. J. (2006). Frames, biases, and rational decision-making in the human brain. *Science, 313*, 684−687.

Deming, A. H. (1989). *Science. Science and other poems* (p. 1994) Baton Rouge: Louisana State University Press.

Descartes, R. (1968). In F. E. Sutcliffe (Ed.), *Discourse on method and the meditations.* London: Penguin Books. (Original work published 1637).

Dewey, J. (1910). *How we think.* Boston: D.C. Heath & Co., Publishers.

Dyson, F. (2006). *The scientist as rebel.* New York: NYREV, Inc.

Elliot, A. J., & Church, M. A. (2003). A motivational analysis of defensive pessimism and self-handicapping. *Journal of Personality, 71*, 369−396.

Emerson, R. W. (2000). In B. Atkinson (Ed.), *The essential writings of Ralph Waldo Emerson.* New York: The Modern Library. (Original work published 1844).

Fischhoff, B., Slovic, P., & Lichtenstein, S. (1977). Knowing with certainty: the appropriateness of extreme confidence. *Journal of Experimental Psychology: Human Perception and Performance, 3*, 552−564.

Fisher, R. A. (1957). Dangers of cigarette-smoking. *British Medical Journal*, 297−298, August 3.

Fisher, R. A. (1958a). Lung cancer and cigarettes? *Nature, 182*, 108.

Fisher, R. A. (1958b). Cancer and smoking. *Nature, 182*, 396.

Fisher, R. A. (1958c). Cigarettes, cancer and statistics. *Centennial Review, 2*, 151−166.

Forbes, V. E., & Calow, P. (2015). Vignette 10.1. Effects of contaminants on population dynamics. In M. C. Newman (Ed.), *Fundamentals of ecotoxicology* (4th ed., pp. 319−324). Boca Raton: CRC/Taylor & Francis.

Forbes, V. E., Calow, P., Grimm, V., Hayashi, T. I., Jager, T., Katholm, A., ... Stillman, R. A. (2011). Adding value to ecological risk assessment with population modeling. *Human and Ecological Risk Assessment, 17*, 287−299.

Forbes, V. E., Calow, P., & Sibly, R. M. (2001). Toxicant impacts on density-limited populations: a critical review of theory, practice, and results. *Environmental Toxicology and Chemistry, 11*, 1249−1257.

Forbes, V. E., Calow, P., & Sibly, R. M. (2008). The extrapolation problem and how population modeling can help. *Environmental Toxicology and Chemistry, 27,* 1987−1994.

Frederick, S., Loewnstein, G. & ÓDonoghue, T. (2002). Time discounting and time preference: A critical review. Journal of Economic Literature, XL, 351−401.

Gamble, T., & Walker, I. (2016). Wearing a bicycle helmet can increase risk taking and sensation seeking in adults. *Psychological Science, 27*(2), 289−294. Available from: http://dx.doi.org/10.1177/0956797615620784.

Gigerenzer, G. (1991). How to make cognitive illusions disappear: Beyond "heuristics and baises. *European Review of Social Psychology, 2,* 83−115.

Gigerenzer, G. (2000). *Adaptive thinking. Rationality in the real world.* Oxford: Oxford University Press.

Gigerenzer, G. (2007). *Gut feelings. The intelligence of the unconscious.* New York: Viking Penguin.

Gigerenzer, G. (2008). *Rationality for mortals.* Oxford: Oxford University Press.

Gigerenzer, G., & Brighton, H. (2009). Homo heuristicus: Why biased minds make better inferences. *Topics in Cognitive Science, 1,* 107−143.

Gigerenzer, G., & Gaissmaier, W. (2011). Heuristic decision making. *Annual Review of Psychology, 62,* 451−482.

Gigerenzer, G., & Goldtein, D. G. (1996). Reasoning the fast and frugal way: models of bounded rationality. *Psychology Review, 103,* 650−669.

Gigerenzer, G., Todd, P. M., & The ABC group. (1999). *Simple heuristics that make us smart.* Oxford: Oxford University Press.

Gilovich, T. (1991). *How we know what isn't so.* New York: Simon & Schuster, Inc.

Gintis, H. (2009). *The bounds of reason.* Princeton: Princeton University Press.

Giraldeau, L.-A., Valone, T. J., & Templeton, J. J. (2002). Potential disadvantages of using socially acquired information. *Philosophical Transactions of the Royal Society of London B, 357,* 1559−1566.

Gould, S. J. (1991). Exaptations: A crucial tool for evolutionary psychology. *Journal of Social Issues, 47,* 43−65.

Grifffith, D. (1999). Exaggerating environmental health risk: The case of the toxic dinoflagellate *Pfiesteria. Human Organization, 58,* 119−127.

Hall, L. W., Jr., Scott, M. C., & Killen, W. D. (1998). Ecological risk assessment of copper and cadmium in surface waters of Chesapeake Bay watershed. *Environmental Toxicology and Chemistry, 17,* 1172−1189.

Hall, L. W., Jr., Scott, M. C., Killen, W. D., & Unger, M. A. (2000). A probabilistic ecological risk assessment of tributyltin in surface waters of the Chesapeake Bay watershed. *Human and Ecological Risk Assessment, 6,* 141−179.

Horton, R. (2003). *Health wars: On the global front lines.* New York: NY: Review Books.

Janis, I. L., & Mann, L. (1977). *Decision making. A psychological analysis of conflict, choice, and commitment.* New York: The Free Press.

Johnson, E. J., & Goldstein, D. (2003). Do defaults save lives? *Science, 302,* 1338−1339.

Jones, E. E., & Berglas, S. (1978). Control of attributions about self through self-handicapping strategies: the appeal to alcohol and the role of underachievement. *Personality and Social Psychology Bulletin, 4,* 200−206.

Kahneman, D., & Tversky, A. (1979). Prospect theory: an analysis of decision under risk. *Econometrica, 47,* 263−291.

Kahneman, D., & Tversky, A. (1996). On the reality of cognitive illusions. *Psychology Review, 103,* 582−591.

Kass, R. E., & Raftery, A. E. (1995). Bayes factors. *Journal of the American Statistical Association, 90,* 773—795.

Kendal, J., Giraldeau, L. A., & Laland, K. (2009). The evolution of social learning rules: Payoff-biased and frequency-dependent biased transmission. *Journal of Theoretical Biology, 260,* 210—219.

Kolata, G. (1984). The proper display of data. *Science, 226,* 156—157.

Kramer, M. S. (1986). Assessing causality of adverse drug reactions: Global introspection and its limitations. *Drug Information Journal, 20,* 433—437.

Kuhn, T. S. (1970). *The structure of scientific revolutions* (2nd ed.). Chicago: The University of Chicago Press.

Kuhn, T. S. (1977). *The essential tension.* Chicago: The University of Chicago.

Lane, D. A., Kramer, M. S., Hutchinson, T. A., Jones, J. K., & Naranjo, C. (1987). The causality assessment of adverse drug reactions using a Bayesian approach. *Journal of Pharmaceutical Medicine, 2,* 265—283.

Liberman, V., Minson, J. A., Bryan, C. J., & Ross, L. (2012). Näive realism and capturing the "wisdom of dyads.". *Journal of Experimental Social Psychology, 48,* 507—512.

Loeb, A. (2010). The right kind of risk. *Nature, 467,* 358.

Mahoney, M. J., & DeMonbreun, B. J. (1977). Psychology of the scientist: an analysis of problem-solving bias. *Cognitive Therapy & Research, 1,* 229—238.

Marubini, E., & Valsecchi, M. G. (1995). *Analyzing survival data for clinical trials and observational studies.* Chichester: John Wiley & Sons, Inc.

Matthews, R. (2000). Storks deliver babies (p = 0.008). *Teaching Statistics, 22,* 36—38.

May, R. M. (1974). Biological populations with nonoverlapping generations: stable points, stable cycles, and chaos. *Science, 186,* 645—647.

McNeil, B. J., Pauker, S. G., Sox, H. C., Jr., & Tversky, A. (1982). On the elicitation of preferences for alternative therapies. *The New England Journal of Medicine, 306,* 1259—1262.

Messerli, F. H. (2012). Chocolate consumption, cognitive function, and Nobel laureates. *The New England Journal of Medicine, 367,* 1562—1564.

Miller, R. G., Jr. (1981). *Survival analysis.* New York: John Wiley & Sons, Inc.

Mynatt, C. R., Doherty, M. E., & Tweney, R. D. (1977). Confirmation bias in a simulated research environment: An experimental study of scientific inference. *Quarterly Journal of Experimental Psychology, 29,* 85—95.

Nelson, R. H. (2004). Environmental religion: A theological critique. *Case Western Reserve Law Review, 55,* 51—80.

Newman, M. C. (2013). *Quantitative ecotoxicology* (2nd ed.). Boca Raton: CRC/Taylor & Francis.

Newman, M. C. (2015). *Fundamentals of ecotoxicology* (4th ed.). Boca Raton: CRC/Taylor & Francis.

Newman, M. C., & Clements, W. H. (2008). *Ecotoxicology. A comprehensive treatment.* Boca Raton: CRC/Taylor & Francis.

Newman, M. C., & Evans, D. A. (2002). Enhancing belief during causal assessments: cognitive idols or Bayes's theorem? In M. C. Newman, M. H. Roberts, & R. C. Hale (Eds.), *Coastal and estuarine risk assessment* (pp. 73—96). Boca Raton, FL: CRC/Lewis Publishers.

Newman, M. C., Zhao, Y., & Carriger, J. F. (2007). Coastal and estuarine ecological risk assessment: The need for a more formal approach to stressor identification. *Hydrobiologia, 577,* 31—40.

Oppenheimer, D. M. (2008). The secret life of fluency. *Trends in Cognitive Sciences, 12,* 237—241.

Osserman, R. (2010). How the Gateway Arch got its shape. *Nexus Network Journal, 12,* 167—189.

Paerl, H. W., Pinckney, J. L., Fear, J. M., & Peierls, B. L. (1998). Ecosystem response to internal and watershed organic matter loading: consequences for hypoxia in the eutrophying Neuse River estuary, North Carolina, USA. *Marine Ecology Progress Series, 166,* 17–25.

Piatelli-Palmarini, M. (1994). In M. Piatelli-Palmarinin, & K. Botsford (Eds.), *Inevitable illusions. How mistakes of reason rule our minds.* New York: John Wiley & Sons, Inc.

Platt, J. R. (1964). Strong inference. *Science, 146,* 347–353.

Popper, K. R. (1956). *Realism and the aim of science.* New York: Routledge. (English trans. 1983).

Rendell, L., Boyd, R., Cownden, D., Enquist, M., Eriksson, K., Feldman, M. W., ... Laland, K. N. (2010). Why copy others? Insights from the social learning strategies tournament. *Science, 328,* 208–213.

Richardson, R. C. (2007). *Evolutionary psychology as maladaptive psychology.* Cambridge: The MIT Press.

Richerson, P. J., & Boyd, R. (2005). *Not by genes alone. How culture transformed human evolution.* Chicago: The University of Chicago Press.

Rizak, S. N., & Hrudey, S. E. (2006). Misinterpretation of drinking water quality monitoring data with implications for risk management. *Environmental Science & Technology, 40,* 5244–5250.

Rogers, E. M. (1995). *Diffusion of innovations* (4th ed.). New York: The Free Press.

Rohr, J. R., & McCoy, K. A. (2010). A qualitative meta-analysis reveals consistent effects of atrazine on freshwater fish and amphibians. *Environmental Health Perspectives, 118,* 20–32.

Samuelson, W., & Zeckhauser, R. (1988). *Status quo* bias in decision making. *Journal of Risk and Uncertainty, 1,* 7–59.

Saris, W. E., Revilla, M., Krosnick, J. A., & Shaeffer, E. M. (2010). Comparing questions with agree/disagree response options to question with item-specific response options. *Survey Research Methods, 4,* 61–79.

Schuldt, J. P., Konrath, S. A., & Schwarz, N. (2011). "Global warming" or "climate change"? Whether the planet is warming depends on question wording. *Public Opinion Quarterly, 75,* 115–124.

Shah, A. K., & Oppenheimer, D. M. (2008). Heuristics made easy: An effort-reduction framework. *Psychology Bulletin, 137,* 207–222.

Spalding, L. R., & Hardin, C. D. (1999). Unconscious unease and self-handicapping. *Psychological Science, 10,* 535–539.

Sprague, J. B. (1969). Measurement of pollutant toxicity to fish. I. Bioassay methods for acute toxicity. *Water Research, 3,* 793–821.

Steele, C. M. (1997). A threat in the air. *How stereotypes shape intellectual identity and performance. American Psychologist, 52,* 613–629.

Strickland, B., & Mercier, H. (2014). Bias neglect: A blind spot in the evaluation of scientific results. *The Quarterly Journal of Experimental Psychology, 67,* 570–580.

Tversky, A., & Kahneman, D. (1974). Judgment under uncertainty: Heuristics and biases. *Science, 185,* 1124–1131.

Tversky, A., & Kahneman, D. (1983). Extensional versus intuitive reasoning: The conjunction fallacy in probability judgment. *Psychological Review, 90,* 293–315.

Verplanken, B., & Holland, R. W. (2002). Motivated decision making: Effects of activation and self-centrality of values on choices and behavior. *Journal of Personality and Social Psychology, 82,* 434–447.

Viscusi, W. K., & Chesson, H. (1999). Hopes and fears: The conflicting effects of risk ambiguity. *Theory and Decision, 47*, 153−178.

Wainer, H. (1984). How to display data badly. *The American Statistician, 38*, 137−147.

Walton, D. (1996). *Arguments from ignorance.* University Park: The Pennsylvania State University Press.

Wason, P. C. (1960). On the failure to eliminate hypotheses in a conceptual task. *The Quarterly Journal of Experimental Psychology, 12*, 129−140.

Weinstein, N. D. (1989). Optimistic biases about personal risks. *Science, 246*, 1232−1233.

Wray, K. B. (2003). Is science really a young man's game? *Social Studies of Science, 33*, 137−149.

Yanis, I. (2004). The benefit of additional opinions. *Current Directions in Psychological Science, 13*, 75−78.

Wheeler, J. R., Grist, E. P. M., Leung, K. M. Y., & Crane, M. (2002). Species sensitivity distributions: data and model choice. *Marine Pollution Bulletin, 45*, 192−202.

Chapter 3

Human Reasoning: Within Scientific Traditions and Rules

3.1 HUMAN REASONING IN A SCIENTIFIC CONTEXT

A skeptical reader might assert that the cognitive errors described in Chapter 2, Human Reasoning: Everyday Heuristics and Foibles, will influence judgments of many ecotoxicologists, but their influence will be limited by adherence to the scientific method. In the absence of contrary evidence, this assertion seems reasonable. This chapter explores the potential influence of cognitive errors when formal scientific methods are employed in the new science of ecotoxicology.

The scientific method is engrained in all grade school youngsters as having five rigid steps: (1) identifying an unresolved problem or question (2) defining a testable hypothesis about that problem, (3) designing an experiment to test the hypothesis, (4) carefully conducting that experiment and analyzing the results, and (5) making evidence-based conclusions about the hypothesis that provide new insight into the original problem or question. In reality, only some (few?) scientists, including ecotoxicologists, consistently conduct business in this manner (Feyerabend, 2010). This being the case, it becomes important to know what are the different ways that legitimate science is conducted and how the foibles described in Chapter 2, Human Reasoning: Everyday Heuristics and Foibles, fared in each.

An illustration involving validation of the increasingly popular species sensitivity distribution (SSD) method highlights the importance of a discussion about how science should be conducted (Fig. 3.1). Before the introduction of the SSD method, there existed a history of pragmatically setting regulatory concentration limits based on some low percentile (e.g., 5% or 10%) of a collection of LC50, EC50 or No Observed Effect Concentration (NOEC) values for several test species. Instead of setting regulatory limits at the values for the most sensitive tested species, the limit was set to a concentration below which only a small percentage of the gathered species metrics would fall. With SSDs, the approach evolved to a point where inferences about the impact on ecological communities were being made from

The Nature and Use of Ecotoxicological Evidence. DOI: https://doi.org/10.1016/B978-0-12-809642-0.00003-0

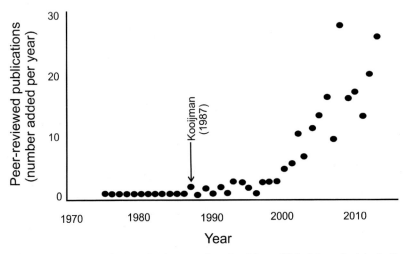

FIGURE 3.1 Number of SSD-related peer-reviewed articles published from its introduction (Kooijman, 1987) through 2013. The interest in the approach as reflected in the annual rate of publication on the topic increased in 1998 after an initial lag period of roughly a decade. Numbers of publications were estimated by conducting literature searches for the phrase, "species sensitivity distribution" with the Web of Science, ProQuest Science & Technology, and Google Scholar databases. The title and abstract of each candidate publication were used to determine article relevance.

predictions (hazardous concentrations, HC_x where x is a percentage) for certain percentage of species from a distribution of species sensitivities. Kooijman (1987) set the statistically defined lower bound of such a prediction as "a lower bound for concentrations that can be expected to be harmful for a given community."[1] Soon after its introduction, inferences from the SSD approach were challenged as being inconsistent with existing ecological concepts (Forbes & Forbes, 1993; Hopkin, 1993). In terms outlined in Chapter 2, Human Reasoning: Everyday Heuristics and Foibles, proponents of the new SSD approach were believed to be victims of segregation error; their activities were done without due consideration of the larger context into which they fit. Again in 1993, an effort to validate single-species NOEC-based SSDs for regulatory purposes concluded, "With reservations, due to [a] paucity of data, it is concluded that single-species toxicity data can be

---

1. When publishing this pivotal SSD paper, Kooijman was employed by a Netherlands organization (TNO) with the stated goal of assuring that scientific research be put to service of society as efficiently as possible. Kooijman's hesitancy about the intended regulatory use of SSD-derived hazardous concentrations was suggested in the paragraph following the quoted material, "All honest scientific research workers will feel rather uncomfortable with such a task, and the author is no exception. The feeling finds its roots in the extrapolation of experimental findings far beyond the limits of our knowledge" (Kooijman, 1987).

used to derive 'safe' values for the aquatic ecosystem" (Emans, Plassche, Canton, Okkerman & Sparenburg, 1993). This tepid validation suggests that confirmation, or perhaps *status quo*, bias might have influenced the authors' interpretation of evidence. Nine years later, Suter (2002) bolstered the perceived validity of the SSD method in his description of its long history of utility, general acceptance, and instances of application by major regulatory agencies. Acceptance of the SSD method by regulatory authorities reinforced its perceived validity via information mimicry (Recollect that, although the mimicry strategy was described in Chapter 2, Human Reasoning: Everyday Heuristics and Foibles, as logical under many circumstances, it carries high risk in situations involving substantial uncertainty as in scientific investigation of a new paradigm or concept. Uncritical reliance on authority also invites confirmation bias and acquiescence.). A few years later, a study focused on SSD application for determining pesticide effects to soil invertebrates (Frampton, Jansch, Scott-Fordsmand, Rombke & Van den Brink, 2006) failed to support the SSD and concluded that "... the data indicate that the uncertainty factor for earthworm acute mortality tests (i.e., 10) does not fully cover the range of earthworm species sensitivities and that acute mortality tests would not provide the most sensitive risk estimate for earthworms in the majority (95%) of cases." Smetanova, Blaha, Liess, Schafer, and Beketov (2014) also found poor agreement between SSD predictions and the results from another established bioindicator method, concluding, "... validation with field data is required to define the appropriate thresholds for SSD predictions." In another recent study (Jesenska, Nemethova & Blaha, 2013), data manipulations and methods were varied to determine SSD robustness for predicting the state of an herbicide-impacted watershed. They assessed the SSD by varying exposure duration for effects metrics, herbicide mixture purity, and the manner in which replicate or outlier data were handled, concluding that the approach was generally robust to these permutations. More recently, successful validation of an SSD mode-of-action model was reported for algal species (Nagai & Taya, 2015).

The cumulative effect of this kind of evidence has fostered sustained interest in and application of SSD methods (Fig. 3.1). Because ecotoxicologists and risk assessors use global introspection to gage scientific validity from evidence like that above, it is difficult to know how much cognitive foibles and careful reasoning swayed the process of SSD validation. None of the SSD publications above had explicitly stated hypotheses in the conventional scientific sense (e.g., tested the hypothesis that the SSD-derived HC_5 consistently predicts the threshold of substantial harm to exposed communities). Nevertheless, this is the conclusion often drawn and used for regulatory decisions.

As illustrated with this SSD example, fundamental differences exist among inquiries intended to establish standard methods for regulatory activities, to develop techniques for environmental studies, or to test scientific hypotheses.[2] Although the above SSD studies addressed different facets of SSD soundness, it is difficult when reading each to ascertain whether a scientific, technical, or practical goal was the touchstone when judging soundness, robustness, or validity. Also unclear was the degree to which cognitive foibles influenced decisions. The question remains open about which of the above reports should, or should not, be treated as presenting scientifically sound evidence during decision-making. Should the argument based on historical utility and regulatory authority carry as much weight as the formal, yet narrowly focused, study of Frampton and co-investigators? It could be similarly argued that the initial publications rejecting the SSD inferences based on established ecological theory (Forbes & Forbes, 1993; Hopkin, 1993) relied too much on scientific authority and asserted untested speculation into the SSD evaluation. Should the inconsistency of the SSD approach with established ecological theory and principles be given more weight than other evidence, including the historical utility of the SSD to regulators? Furthermore, is SSD robustness to methodological changes a reliable indicator of logical validity? How much did the desire to establish an easily applied technology influence conclusions of Nagai & Taya? Which studies would be included as sound scientific evidence if one were to do an SSD meta-analysis? This chapter describes the scientific context for answering questions of this nature in ecotoxicology.

3.2 TRADITIONS IN SCIENCE

There is no longer any consensus on what the scientific method is.

Hacking (1996)

The idea of a method that contains firm, unchanging, and absolutely binding principles for conducting the business of science meets considerable difficulty when confronted with the results of historical research. We find, then, that there is no one rule, however plausible, and however firmly grounded in epistemology, that is not violated at some time or other.

Feyerabend (2010).

Two classic books, Aristotle's *Organon* collection and Bacon's 1620 *Novum Organum* criticism of and embellishments on Aristotle's methods, will serve here as early milestones in the evolution of western scientific thinking. The discussion will then move to 19th and 20th century milestones

---

2. Ecotoxicology has three major goals (Newman, 1996). Its scientific goal is to generate, organize, and classify knowledge about the fate and effects of toxicants in ecosystems based on testable, explanatory principles. The second is the technical goal of developing and applying tools and procedures to improve our understanding of toxicant fate and effects in ecosystems. Its practical goal is to apply available knowledge, tools, and procedures to solve specific problems.

including falsification in logical and statistical senses. Pertinent features of Kuhn's historical vantage on science (Kuhn, 1962), strong inference (Platt, 1964) including the method of multiple working hypotheses (Chamberlin, 1897), and abductive inference will be the final milestones.

3.2.1 Historical Wisdom

Aristotle's approach to scientific inquiry involved observation, expression of a general principle or theory to explain observations, and then deductions about the phenomena based on that ruling theory (Losee, 2001). Observation to Aristotle meant evidence from sense experience, not experimental observation involving manipulations of conditions. Inferences began with induction from observations to some general explanatory principle which is subsequently considered to be true. Then other observations or properties were explained deductively from that ruling theory or principle. For example, one can apply induction that a chemical with a particular substructure is a poison because several chemical compounds with that substructure were observed to kill organisms. Presented with an untested compound with the same substructure, one can deduce from the newly established principle that it would also be poisonous. One observes using the senses, generalizes to an explanation by induction, and then makes specific deductions from this general explanation. This ancient advance in how nature is explored was very influential on the developing sciences of medieval Europe after Aristotle's work was eventually translated from Greek to Latin. Regrettably, his approach remained susceptible to many of the foibles described in Chapter 2, Human Reasoning: Everyday Heuristics and Foibles. It was prone to precipitate explanation that is, the invoking of a general principle or ruling theory as sufficient explanation for evidence without the obligation to further test the adequacy of that explanation

During the 17th century, Francis Bacon criticized the medieval scientific methods stemming from the *Organon*. He saw them as nonsystematic and inclined to lead its practitioners to generalize too quickly from evidence. Once the ruling theory was established as being true, negative instances were not actively sought to aid in any further inferences. As applied at the beginning of the Renaissance, such an approach was too error-prone to remain the foundation for modern science. To argue this point, Bacon compiled errors ("idols")[3] made in related inferences including several which we discussed

---

3. Bacon defined four kinds of idols: idols of the tribe (errors inherent to the human mind), the den (errors due to the nature of an individual such as those described earlier under Section 2.2.2), the market (errors emerging during social interaction as discussed in Chapters 6–8), and the theater (errors arising from dogmas of different schools, disciplines, as will be referred to soon when discussing the evolution of sciences).

already as cognitive errors. Here are pertinent idols of the tribe from Bacon's *Novum Organum* (Bacon, 1620).

- "The human understanding is, by its own nature, prone to abstraction, and supposes that which is fluctuating to be fixed. . ."
- "The human understanding, from its peculiar nature, easily supposes a greater degree of order and equality in things than it really finds . . ..will invent parallels and conjugates and relatives, where no such thing is" (see Section 2.2.2.12)
- "The human understanding, when any proposition has been once laid down (either from general admission and belief, or from the pleasure it affords), forces everything else to add fresh support and confirmation; and although most cogent and abundant instances may exist to the contrary, yet either does not observe or despises them, or rid of and rejects them by some distinction, with violent and injurious prejudice, rather than sacrifice the authority of its first conclusions" (see Sections 2.2.2.5 and 2.2.2.6)
- "The human understanding is most excited by that which strikes and enters the mind at once and suddenly, and by which the imagination is immediately filled and inflated" (see Sections 2.2.2.2−2.2.2.4)
- "The human understanding resembles not dry light but admits a tincture of the will and passions, which generate their own system accordingly; for man always believes more readily that which he prefers" (see Section 2.2.2.2−2.2.2.6 and 2.2.2.11)

Bacon's modifications to the scientific approach aimed to methodically reduce the influence of these idols (cognitive and social foibles). Although thought at the time to be unnecessary and impractical, Bacon's modifications gradually gained acceptance by scientists who began to make inferences with evidence from structured experiments in addition to unstructured observations of nature. This inductive process was carried out through a progression of tests and the gradual amassing of evidence that incrementally added or detracted credibility from a candidate explanation. Explanations were no longer considered true or false: they were possible explanations subject to rejection if future investigations failed to support them. Importantly, Bacon highlighted the value of the exclusion for culling away spurious or false relationships.[4] Also key to Bacon's approach was the value placed on critical testing ("fingerpost" instances) that resulted in a clear swing of logical support away from one hypothesis to another competing hypothesis (Losee, 2001). Critical, especially exclusionary, tests were superior to weakly discriminating tests in scientific inference. Many weak tests that heretofore

---

4. To give Aristotle credit, he also did discuss the value of negation versus affirmation but his treatment in the *Organon* (specifically in Chapter 1: the prior analytics) seems more focused on judging syllogisms than nature itself.

were convincing based on repetition alone (the foible discussed in Chapter 2, Human Reasoning: Everyday Heuristics and Foibles, as availability as influenced by familiarity) were now viewed as less useful than a single critical test.

William Whewell, during the 19th century, took a historical view of science practices and proposed the process of melding evidence and explanations that is currently called the hypothetico-deductive (H-D) method. In Whewell's method, facts are gathered and ideas used in a particular way to explain them. First, facts are gathered, analyzed, and more collected if required to move to the next step. Second, a pattern is suggested by the data, producing a possible explanation (i.e., an "idea" in Whewell's terms). Finally, the evidence and explanation are further integrated and enriched. In Whewell's H-D method, facts accumulate until induction allows the establishment of an explanatory theory. Whewell found that no particular set of rules was always present during this three-step process (Losee, 2001).

At this 19th century stage of science, heavy reliance on the H-D method emerged in which a theory or explanation's survival of repeated testing enhanced its credibility. Multiple lines of evidence were brought together to formulate a robust explanation. In contrast to Aristotle's ruling theory, explanations were conceptualized as working hypotheses that might be proven wrong (falsified) as evidence accumulated. An explanation might eventually be considered conditionally confirmed if it survived repeated, rigorous testing. An exemplary form of scientific H-D inference is explored further in Section 3.2.2. Unfortunately, the rigors of testing and the number of tests required to corroborate an explanation or theory remain open to interpretation with this approach.

Recurrent reliance on induction forced thoughtful scientists to confront "the problem of induction." Recollect that inductive reasoning starts with a general theory or explanation and then makes predictions from it. Hume's classic problem of induction is usually explained with an example such as the following. You propose that all crows are black and then observe many black crows. However, no matter how many crows you survey, there is a chance that the next crow will be another color. This is the problem of induction that makes induction alone a compromised tool for scientific reasoning. A partial solution to the induction problem was introduced by Karl Popper. He reasoned that one can use evidence to falsify a proposition or theory but not to logically validate it. Building upon Bacon's emphasis on efficient tests that excluded other explanations, he stressed, "Thus we shall take the greatest interest in the falsifying experiment" (Popper, 1959). Popper also favored hypotheses formulated in quantitative terms because they were more explicit and consequently strong falsifying tests were more easily developed for them than for qualitative hypotheses. Greatly influenced by the logical clout of falsification, Ronald Fisher developed at the beginning of the 20th century the currently most prominent form of statistical inference. Although Fisher's null hypothesis significance test (NHST) is still considered the gold standard for inference in many sciences, results from these tests are pervasively misunderstood as being capable of confirming hypotheses (see Chapter 5: Individual Scientist: Reasoning by the Numbers).

The current *status quo* in many scientific disciplines, including ecotoxicology, is a statistical inference from evidence generated by careful observation or structured experiments. Recent tradition relies heavily on falsification-based inference although violations of the associated logic are commonplace. As now presented to the advanced science student, the modern scientific method involves the following steps: (1) identifying an unresolved problem or question, (2) stating a conceptual hypothesis that might contribute insight about the question, (3) based on the conceptual model, stating a testable statistical null hypothesis that can potentially be falsified and used to make inferences about the conceptual hypothesis, (4) designing an experiment to test that statistical hypothesis, (5) conducting the experiment and analyzing the results using the NHST, and (6) making evidence-based conclusions about the conceptual hypothesis from the NHST outcome. Not much can be concluded if the null hypothesis is not rejected (falsified), so most researchers conduct their research in hopes that it will be rejected (Biau, Jolles & Porcher, 2010). If the null hypothesis is rejected, inferences are made via the H-D method based on the statistical alternative hypothesis. (Fundamental errors in this conventional approach will be discussed in Chapter 5, Individual Scientist: Reasoning by the Numbers.) A community of scientists independently investigating the conceptual hypothesis via the statistical falsification scheme eventually will assign an evidence-based level of credibility to it. Each scientist in the community is assumed to carry their own biases, yet a collective intersubjective testing within the scientific community is believed to reduce the impact of individual subjectivities (Popper, 1959).

A career-long intellectual challenger of Karl Popper, Thomas Kuhn (1962) countered that Popper's orderly scientific process was not what practicing scientists actually do. He grossly divided scientists' practices into two categories: normal and innovative science. Some scientists had a proclivity for one or the other, whereas others moved between the two as required by their investigations. Nonetheless, a balance of both was thought to be essential in any healthy science. Normal scientists work within a currently accepted core theory (paradigm), enriching detail and accuracy of associated measurements.(On page viii of *The Structure of Scientific Revolutions*, Kuhn (1962) defines a scientific paradigm. "These I take to be universally recognized scientific achievements that for a time provide model problems and solutions to a community of [science] practitioners.") Normal scientists do not intend to test that paradigm and pragmatically accept it as valid in order to ply their trade. They measure, model, and adjust satellite features and auxiliary hypotheses surrounding the core paradigm. As normal science adds and refines details, exceptions and inexplicable evidence emerge that draws suspicion about the adequacy of the core explanatory paradigm. Innovative science comes into play when enough mismatches between accumulating evidence and the paradigm create a crisis (Kuhn, 1962). In contrast to normal scientists, a scientist practicing innovative science intends to test existing theory and perhaps propose an alternative.[5] The testing of innovative scientists causes a revolution with the established paradigm being eventually displaced by one that better explains the current body of evidence. So, the activities of normal scientists reveal anomalies and those of innovative scientists then establish a new paradigm that better explains existing evidence. The process is recursive with no currently favored paradigm assumed to be capable of surviving indefinitely.

Kuhn asserted that no definitive explanations are ever found by scientists using any prescribed approach. No specific scientific process was capable of uncovering definitive explanations of nature. Explanations in the form of theories—Kuhn's paradigms—evolve through time as knowledge accumulates through normal science and the scientific community's paradigms shift via innovative science. Paradigms were not static, and any explanation favored at a particular time is expected to be succeeded by another as the science evolves. "Paradigms gain their status because they are more successful than their competitors in solving a few problems that the group of practitioners has come to recognize as acute" (Kuhn, 1962).

The scientific method as described to this point has likely become less concrete than many readers thought. Perhaps, due to the leveling and sharpening expected with retelling through time, most of us were led to believe that the scientific method was an inviolate set of principles discovered long ago by inspired scientific visionaries. Any scientist who did not adhere to The Scientific Method was poorly trained or professionally uncouth. Nothing could be further from the truth. The scientific method encompasses several approaches and continues to evolve. For example, it is my opinion that the falsification context for scientific investigation will be replaced by quantitative abductive methods during this century (see Chapter 5: Individual Scientist: Reasoning by the Numbers).

Despite the plurality of methods in science, the majority of readers would still recognize flawed science when confronted by it. Although there is no one scientific method generally agreed upon, norms seem to exist for

---

5. Predictably, the resulting shifts required by a crisis can be unsettling to scientists whose reputation and time spent studying the paradigm are being threatened. I suspect that Kuhn picked the emotion-laden word, crisis, for this reason. Overreactions to a crisis can foster unobjective opiniatrety, and in some cases, pathological science as described in Chapter 4, Pathological Reasoning Within Sciences.

scientific soundness. This begs the question, "what are the distinguishing norms of acceptable scientific inquiry and inference?"

In my opinion, the distinctive, common feature of legitimate scientific approaches is that they all endeavor to objectively construct explanations from observations, or preferably experimental evidence, in a way that is open to scrutiny and challenge by others in the scientific community. An essential norm is minimum subjectivity and maximum objectivity of practices and interpretations of evidence. Practitioners reduce intersubjective bias by applying logical tools and external peer review. The precipitate explanation was acceptable in the early Renaissance period but is now considered invalid.

What a sophisticated reviewer weights most heavily in a scientific manuscript is the soundness and reproducibility of a normal scientist's evidence, the evidence-based plausibility of an innovative scientist's proposed explanation or theory, and reasonableness of conclusions drawn from the evidence. Also essential is convincing proof that the researcher and results were acceptably objective. Unavoidably, the reviewer's judgments will be colored by contemporary norms and expectations that change through time. A new analytical tool developed by the normal scientists might be suspect to a reviewer simply because it is unfamiliar. Acceptance requires a detailed description by the normal scientists to avoid rejection. Likewise, a novel theory from an innovative scientist might be viewed with suspicion unless the reviewer is presented with compelling and balanced reasons for doubting an existing explanatory theory and favoring the new one.

The reader may be discomforted by the above narrative that replaces a formulaic scientific method with a more fluid one. It might leave the false impression that the process is more forgiving than first taught at school and consequently more prone to shoddy science. To dispel such an impression, two exemplary vantages (classic strong inference and abductive inference) are detailed below.

3.2.2 Falsification and Strong Inference

Hopefully, the reader accepts the proposition in Chapter 1, The Emerging Importance of Pollution, that the science of ecotoxicology must advance as efficiently as possible in the 21st century. Ecotoxicology has an historic obligation to society to provide sound insights with which to manage emerging environmental problems. This requires the application of the most effective scientific approaches. To facilitate understanding of the best approaches, the classic publication, *Strong Inference* (Platt, 1964) that addresses the question of what scientific approaches accelerate knowledge growth of a science, will be discussed. Integrated into Platt's strong inference is the method of multiple working hypotheses (Chamberlin, 1897), which will also be detailed.

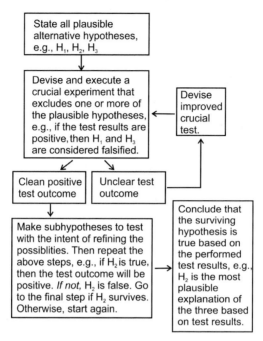

FIGURE 3.2   Flow diagram of Platt's strong inference. See text for detailed explanation.

Platt's classic *Science* article begins with the observation that all sciences do not progress at the same rate. He then proposes a reason for the high rate of progress in some sciences and the lack of progress in others. The distinguishing trait is "applying the [steps in Fig. 3.2.] to every problem in science, formally and explicitly and regularly..." (Platt, 1964). The net result of doing so within a discipline is strong inference. A quick glance at Fig. 3.2 would suggest that strong inference is just the conventional scientific method (*sans* statistics) described above as the currently accepted method in many disciplines; however, there are two distinctive features. Most important is the expectation that strong inference will be used formally, explicitly, and regularly by all members of that discipline. This is central to fostering rapid progress in the field. Researchers veering from this method will generate weak inferences that muddle the literature about the plausibility of candidate explanations or theories. He likens this to trying to get across a lake with a boat motor having some cylinders that only fire occasionally. "If our motorboat engines were as erratic as our deliberate intellectual efforts, most of us would not get home for supper." The other distinguishing feature of rapidly progressing science is prominent in the first step of strong inference, that is, the stating of all plausible alternative hypotheses (Fig. 3.2). This is the initial step of the method of multiple working hypotheses which Platt borrowed intact from Thomas Chamberlin (1897).

Chamberlin (1897) was already famous for his work in geology when he proposed the method of multiple working hypotheses.[6] Indeed, the method emerged from his experiences trying to explain existing geological features despite high uncertainty and several plausible explanations. He begins his explanation by reminding the reader that Aristotle's ruling theory leads to a precipitate explanation. Facts accumulate and are explained by some ruling theory with a minimal obligation to consider other explanations. Because of the unreliability of this method, it was replaced in history by the working hypothesis which serves as a framework for most modern investigations. A working hypothesis is not considered "true" in a logical sense: it is conditionally accepted as the best current hypothesis and might be replaced should a better hypothesis appear. Chamberlin then extends this narrative to highlight a flaw in the working hypothesis approach.

> *Conscientiously followed, the method of the working hypothesis is a marked improvement upon the method of the ruling theory; but it has defects—defects which are perhaps best expressed by the ease with which the hypothesis becomes a controlling idea. To guard against this, the method of multiple working hypotheses is urged. It differs from the former method in the multiple character of its genetic conceptions and of its tentative interpretations. It is directed against the radical defect of the two other methods; namely, the partiality of intellectual parentage. The effort is to bring up into view every rational explanation of new phenomena, and to develop every tenable hypothesis respecting their cause and history.*
>
> Chamberlin (1897)

Chamberlin's method forces the investigator to put equal effort into ascertaining the mettle of all plausible hypotheses. In so doing, the tendency to feel ownership of an explanation or become vested in a theory is lessened. Instead, the investigator becomes vested in the insightful and incisive manner with which they conduct their inferences.

Platt's strong inference involves the conscientious use of the H-D approach in combination with the method of multiple working hypotheses. The investigator in a field practicing strong inference risks their reputation only if they publish weak inferences where strong inferences were possible.

Strong inference is certainly an improvement over less formal scientific approaches, including inferences based on a single working hypothesis or

---

6. When his vested interests in conventional concepts of geology were threatened, he also became infamous for his aggressive rejection of the now-accepted continental drift theory. The point of mentioning his shortcoming is to emphasize a major theme in this book that even the best and most innovative scientists are susceptible to the errors outlined in Chapters 2 and 4 through 6.

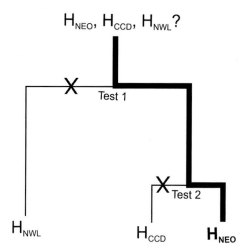

FIGURE 3.3   An example of applying Platt's strong inference approach. In a certain year, 50% of 750 honey bee hives failed and three hypothetical reasons were proposed, H_{NEO}, H_{CCD}, and H_{NWL} (see text for details). Following the method of strong inference, two exclusionary tests (Tests 1 and 2) of the hypotheses are conducted sequentially to identify H_{NEO} as the most likely explanation. Each test resulted in a rejection decision, leading the investigator through a dichotomous decision tree to the most likely choice.

rote application of a single NHST. But a simple, fictional example can illustrate the limitations of Platt's strong inference approach.[7]

One year in an agricultural area dependent on bee pollination, an estimated 50% of 750 established honeybee hives failed. To decide on a course of action, farmers must judge whether the failure was caused by normal winter loss (H_{NWL}), colony collapse disorder (H_{CCD}), or neonicotinoid pesticide application (H_{NEO}). As shown in Fig. 3.3, a significance test might reject the hypothesis of normal winter loss (H_{NWL}), leaving only two hypotheses as plausible. A second test might reject H_{CCD} by gauging whether the level of colony collapse disorder (CCD) was significantly higher than that of past years. The elimination of H_{CCD} and test survival by H_{NEO} leads to the inference that spraying with neonicotinoid pesticides produced the high level of hive failures.

---

7. Logical falsification with a Fisherian significance test has been described to this point. Such significance tests attempt to nullify one hypothesis, the null hypothesis, and consequently have only one error type, Type I. It is expressed as the probability of falsely rejecting the null hypothesis and is set at an acceptable rate (α) prior to testing. However, this example will use the Neyman–Pearson hypothesis testing approach that has an additional error, the Type II error, the probability of falsely rejecting the alternative hypothesis. Unlike Fisher's significance testing, hypothesis testing involves two hypotheses and decision errors. Also, a hypothesis test does not logically falsify a hypothesis as attempted by significance testing. It is designed to aid the scientist in deciding which is the most probable of alternative hypotheses. More discussion of this will be provided in Chapter 5: Individual Scientist: Reasoning by the Numbers.

The adequacy of the above "Accept/Reject" context of strong inference can be further scrutinized by adding more detail about the applied tests. Assume that the normal winter loss in this area, especially due to colony weakening by *Varroa* mites, is 30%. Also assume that to determine if the failure rate of 50% was unusually high with the first test, the local extension service apiarist applied a statistical test with Type I (α) and II (β) error rates of 0.05 and 0.20, respectively. The hypothesis that the failure reflected normal winter loss was rejected and supported the alternative hypothesis that failure was due to another reason such as CCD or neonicotinoid application. This statistical inference is flawed based on widespread confusion about what a *P*-value reflects. The *P*-value is the probability of getting the observed, or more extreme, data over the long run: it is not the probability of the alternative hypothesis being true. The probability of the alternative hypothesis being true given a positive test is called the positive predictive value (PPV) and is calculated from the Type I (α) and II (β) error rates, and a prior (*R*). The *R* is estimated before the test is conducted by dividing the number of "true relationships" by the number of "no relationships" derived by some means such as a literature survey of similar studies. If no such information is available, the principle of indifference would suggest using an estimate of $R = 1.00$.

$$PPV = \frac{(1 - \beta)R}{R - \beta R + \alpha}$$

The PPV for the first test can be estimated assuming values of 0.05 and 0.20 for α and β, respectively, and $R = 1.00$.

$$PPV_{Test\ 1} = \frac{(1 - \beta)R}{R - \beta R + \alpha} = \frac{(1 - 0.20)(1.00)}{1.00 - (0.20)(1.00) + 0.05} = 0.94$$

In contrast to the certitude implicit in the Accept/Reject or True/False dichotomous decision tree in Fig. 3.3, the calculated probability of 0.94 provides slightly less definitive support of any decision that the level of hive failures was not normal winter loss. Also useful would be the probability of the null hypothesis being true if the test had been negative, that is, the negative predictive value (NPV).

$$NPV_{Test\ 1} = \frac{1 - \alpha}{1 + \beta R - \alpha} = \frac{1 - 0.05}{1 + (0.20)(1.00) - 0.05} = 0.83$$

A negative test would have been less informative than a positive test because the NPV is materially lower than the PPV. Clearly, these important deviations from certitude are missing in Platt's decision tree approach.

Perhaps influenced by the certainty effect discussed in Chapter 2, Human Reasoning: Everyday Heuristics and Foibles, the extension agent accepts the positive results of the first hypothesis test as definitive, concluding that excessive hive failure occurred. The agent next hypothesizes that the excessive failure might have resulted from CCD, a hive disorder in which the queen is abandoned by the worker bees. Past studies indicate that the region has a CCD background rate of 25%. The observed level of failed or failing hives due to CCD during the year in question was estimated to be 37%. A statistical test is used to test whether the CCD-associated hive failings were higher than expected and the null hypothesis of no difference was rejected. The Type I and II error rates were 0.05 and 0.35, respectively. The higher Type II error rate was associated with difficulties in scoring a hive as subject to CCD. The PPV and NPV can be estimated if one assumes as above that $R = 1.00$,

$$PPV_{Test\ 2} = \frac{(1 - \beta)R}{R - \beta R + \alpha} = \frac{(1 - 0.35)(1.00)}{1.00 - (0.35)(1.00) + 0.05} = 0.93.$$

$$NPV_{Test\ 2} = \frac{1 - \alpha}{1 + \beta R - \alpha} = \frac{1 - 0.05}{1 + (0.35)(1.00) - 0.05} = 0.73$$

Again, a positive test provides strong, but not definitive, support for the alternative hypothesis. If the testing had failed to reject the null hypothesis, the low NPV indicates weak inference that the null hypothesis is supported by the data. For the reader uncomfortable with these computations, Fig. 3.4

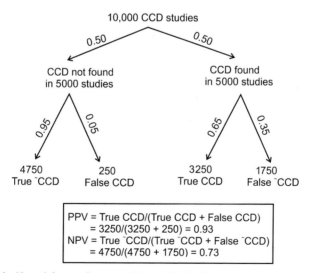

FIGURE 3.4   Natural frequencies approach to estimating PPV and NPV for Test 2. Although the actual number of CCD studies could have been used, the calculations were done using 10,000 CCD studies, so that the example calculations did not involve fractional numbers of hives. The high PPV (0.93) indicates that the significant test outcome was more definitive than if a negative test outcome had occurred (NPV = 0.73). In reality, any decision to move one way or another at a decision branch such as those in Fig. 3.3 involves different levels of credibility.

illustrates how the same estimated PPV and NPV can be approximated using natural frequencies.

To continue the example, an informed estimate of R might be used if there were information in archived extension agency reports indicating that one in five local studies initially claiming CCD as the cause for an outbreak of hive failures was actually shown later to have correctly identified CCD as the cause.

$$\text{PPV}_{\text{Test 2}} = \frac{(1-\beta)R}{R - \beta R + \alpha} = \frac{(1-0.35)(1/4)}{1/4 - (0.35)(1/4) + 0.05} = 0.76$$

$$\text{NPV}_{\text{Test 2}} = \frac{1-\alpha}{1 + \beta R - \alpha} = \frac{1-0.05}{1 + (0.35)(0.25) - 0.05} = 0.92$$

In this instance, the most uncertain inferences would be from a positive test, supporting H_{NEO}. Clearly, the specifics of a statistical test (i.e., α, β, and R) dictate how well results conform to Platt's simple falsification scheme. His scheme does lull the user into thinking in terms of certitudes when probabilities of outcomes might be more appropriate. Platt's original approach becomes difficult to apply in the presence of high uncertainty, non-trivial test error rates, or multiple lines of evidence. Like the *argumentum ad ignorantiam* discussed in Chapter 2, Human Reasoning: Everyday Heuristics and Foibles, the application of Platt's simple scheme carries the obligation to conduct more extensive studies allowing stronger inferences before making any final judgment. In the hive failure example, this might have included defining the errors involved in scoring a hive as failed or failing due to CCD and also examining specific biomarkers of queen health notionally impacted by neonicotinoid exposure. Further studies of queen health biomarkers might also help inferences about H_{NEO} if another plausible hypothesis for the high hive failure level emerged during deliberations.

> *It is occasionally possible to encapsulate a method of science as a recipe. The most satisfying is that based on multiple competing hypotheses, also known as strong inference. It works only on relatively simple processes under restricted circumstances and particularly in physics and chemistry, where context and history are unlikely to affect the outcome.*
>
> Wilson (1998)

3.2.3 Abductive Inference: Strongest Possible Inference

> *It is therefore worth while to search out the bounds between opinion and knowledge.*
>
> Locke (1690)

Platt's strong inference is a highly advantageous way to conduct a scientific inquiry when possible. It can produce adequate inferences alone in some cases, and where this is not the case, it serves as a powerful prelude to a more quantitative inquiry. The accumulation of sound knowledge would certainly be accelerated in our science if every ecotoxicologist aspired to appropriately apply strong inference.

As suggested above, there are instances in which a more explicit approach might be required. Earlier, I (Newman & Clements, 2008) referred to one such approach that extends Platt's strong inference to that of making the strongest possible inference. It is a straightforward modification of the strong inference approach that replaces the logical True/False dichotomous logic with abductive inference. Abductive inference is the inference that favors the most probable explanation or hypothesis: the evidence-based probability of a hypothesis or explanation reflects its plausibility. Embedded in Bayesian statistics, abduction is a well-established context in many fields such as medical diagnostics and artificial intelligence (Josephson & Josephson, 1996) that could substantially improve judgments of ecotoxicologists and ecological risk assessors.

An extension of Chapter 2's, Human Reasoning: Everyday Heuristics and Foibles, opening example of misjudging the presence of a pesticide in a water sample illustrates abductive inference. In that example, a tap water sample tested positive for the presence of a pesticide. Based on a prior probability[8] of the pesticide's presence (0.001) and test error rates (0.05 and 0.02 probabilities of a false-negative or false-positive test, respectively), the posterior probability of the pesticide being present in the samples was calculated (Following standard notation, a superscripted $\sim$ before a hypothesis or state is its negation. For example, here $\sim$Pesticide means "pesticide not present.").

$$P\left(\text{Pesticide}|\text{Test}^+\right) = \frac{P(\text{Pesticide})P(\text{Test}^+|\text{Pesticide})}{P(\text{Pesticide})P(\text{Test}^+) + P(\sim\text{Pesticide})P(\text{Test}^+|\sim\text{Pesticide})}$$

$$P\left(\text{Pesticide}|\text{Test}^+\right) = \frac{(0.001)(0.95)}{(0.001)(0.95) + (0.999)(0.02)} = 0.0454$$

This is not very convincing evidence; but, because of the seriousness of drinking water safety, it might still prompt further scrutiny. Assume that a positive test results for a follow-up sample taken the next week from the suspect tap. The above posterior probability of 0.0454 is used as a prior probability in the above equations to calculate a new posterior probability.

$$P\left(\text{Pesticide}|\text{Test}^+\right) = \frac{(0.0454)(0.95)}{(0.0454)(0.95) + (0.9546)(0.02)} = 0.6932$$

---

8. The terms "prior and posterior probabilities" will appear periodically throughout this book. They refer to the probabilities before the evidence is considered (prior) and after (posterior) it is included in estimating a probability.

This new evidence-based probability adds substantially to the hypothesis that the pesticide is truly present in the tap water. A third sample is taken and results in another positive test. The 0.6932 becomes the new prior probability in the equation.

$$P\left(\text{Pesticide}|\text{Test}^+\right) = \frac{(0.6932)(0.95)}{(0.6932)(0.95) + (0.3068)(0.02)} = 0.9908$$

Based on the updated evidence and abductive inference, the best judgment would be that the pesticide is present and action is warranted. Natural frequency diagrams illustrated previously also can be used. A similar approach with slightly modified equations would have been used if the follow-up tests were negative for pesticide presence. Obviously, this approach would be far superior to the failed application of global introspection described in Chapter 2, Human Reasoning: Everyday Heuristics and Foibles.

More involved applications of abductive inference in ecotoxicology are presented in Newman, Zhao, and Carriger (2007), chapter 36 of Newman & Clements (2008), Carriger & Newman (2011), and chapter 1 of Newman (2013), including some based on natural frequency diagrams and Bayesian networks. Citations in these sources also apply abduction to similar problems in other disciplines. As an example, illustrations for judging the probability of an adverse drug reaction (i.e., Lane, 1989; Lane, Kramer, Huthinson, Jones & Naranjo, 1987) can easily be adapted to ecotoxicological concerns. The following, final example in this chapter will recast example 1 in Lane (1989) to demonstrate how different kinds of evidence can be handled with quantitative abductive methods.

Assume that a small population of an endangered waterfowl nests in a pond with elevated PAH sediment concentrations. A chemical spill occurs and you must judge whether the spill or chronic PAH exposure caused the death of a chick. Past studies estimated that 1 in 10 chicks dies due to the PAH contamination during a critical 48-day post-hatch period. A truck spill of Toxicant X occurs, resulting in toxic concentrations for 1 day before volatilization brings them down again to a safe concentration. The chance of a chick dying from exposure to that toxicant concentration is estimated to be 1 in 40. A chick actually dies on the day of the spill, and the responsible agency must decide if enough evidence exists to claim that the spilt toxicant was the likely cause of death.

Two sorts of evidence need to be combined: the probability of death with exposure to PAH or toxicant X and the timing of the chick death. The probability of a chick dying from exposure to the Toxicant X is 1/40 and that of a chick dying due to PAH exposure is 1/10, or making 40 the common denominator, 4/40. So, for every five deaths during the 48-day period, PAH exposure would have caused four and Toxicant X would have caused one. Now timing needs to be integrated into the calculations. A chick succumbing to PAH could have died any day in the 48-day period, so a PAH-poisoned chick's probability of dying on the day of the spill is 1/48. Obviously, death from the spill could only occur on 1 day, so the probability of a chick that died from the spill dying during the spill day is 1/1. With this evidence, the ratio of probabilities of dying from these two candidate causes can be estimated.

$$\frac{P\left(\text{Death}_{\text{Toxicant X}}|\text{Death}_{\text{Background}}, \text{Timing}\right)}{P\left(\text{Death}_{\text{PAH}}|\text{Death}_{\text{Background}}, \text{Timing}\right)}$$

$$= \frac{P\left(\text{Death}_{\text{Toxicant X}}|\text{Death}_{\text{Background}}\right)}{P\left(\text{Death}_{\text{PAH}}|\text{Death}_{\text{Background}}\right)}$$

$$x \frac{P\left(\text{Timing}|\text{Death}_{\text{Toxicant X}}, \text{Death}_{\text{Background}}\right)}{P\left(\text{Timing}|\text{Death}_{\text{PAH}}, \text{Death}_{\text{Background}}\right)}$$

$$= \frac{1/5}{4/5} \times \frac{1/1}{1/48} = 12$$

The ratio of probabilities (12:1 or 12/13 = 0.92%) indicates that the spill was the most likely cause of the chick's death. There is enough evidence to charge the trucking company; however, the judge hearing the case might be more easily convinced by a diagram of natural frequencies than the above equation.

3.2.4 Weaker Forms of Inference

The strongest inference methods result in optimal decisions when applied thoughtfully to existing evidence. There are instances in which their application is impossible or so conditional that they might distract the researcher from the real question. This might be the case during many initial studies into a new research question because insufficient evidence is available to conduct rigorous testing. On the other hand, any inferences from a simple application of Platt's strong inference will not have the computed probabilities of making the right choice at each branch in the decision tree. Strong inference would still produce extremely valuable insight, but in the best of all worlds, would be followed by studies permitting the calculation of probabilities. Also, straightforward abductive inference as illustrated might not be advisable if the assumptions underpinning the probability calculations or applied techniques are not met (Bollen & Pearl, 2013). For instance, effects from joint exposure to Toxicant X and PAH might not be additive.

Inferences in ecotoxicology and ecological risk assessment often have "weight-of-evidence" (WOE) as their foundation. As just mentioned, this might be the only choice available, but if used, it is important to remember that WOE is often just another way of saying global introspection (see Section 2.2.2.22). Our previous discussions highlighted the high potential of weak or incorrect inferences based on informal global introspection.

Rule-based WOE is a particular form of WOE applied during many ecological risk assessments. It relies on a set of qualitative rules for gathering and weighing evidence to make causal inferences. The best-known set of such rules is that proposed by Hill (1965) for examining the association or causation of occupational diseases. Hill's aspects that enhance the credibility of an association between an agent and disease can be summarized as follows.

- A strong association
- Consistency of an observed association
- Specificity of an association
- Consistent temporal sequence (cause present as a precondition to seeing the disease or high incidence of the disease)
- A biological gradient (higher amounts of an agent produce higher levels or chance of an effect)
- Existence of a plausible biological mechanism
- Coherence with our general knowledge base
- Especially, the presence of experimental evidence
- Analogy drawn from another disease-causing agent

Several other rule sets such as those of Evans (1976), Beyers (1998), Fox (1991), and Suter (1993) are derivative of either Hill's aspects or Koch's postulates. Still others such as Last (1983) or Green & Byar (1984) are used to classify the quality of epidemiological evidence. The weakest epidemiological evidence comes from anecdotal reports of adverse effects/disease. Better evidence comes from case series without controls: even better evidence emerges from case series drawing on literature controls. Stronger still are inferences from statistical analyses involving subsets of harmed individuals who were or were not exposed to the suspected disease agent. The most powerful evidence comes from controlled experiments or trials.

All of these rules for WOE judgments can be helpful but are more susceptible to cognitive errors and social influences than are those from strong inference and abductive inference. For example, it was the inferential weakness of Hill's aspects that allowed Ronald Fisher and later the tobacco industry to dismiss the proposed association between tobacco use and disease.[9] Fortunately, Hill's proposed disease linkage stimulated further research that eventually provided strong support for the association.

---

9. The exchange between Fisher, an avid pipe smoker, and Hill was mentioned earlier (Section 2.2.2.10) in the context of the value centrality bias. Hill's (1965) comment "... I think we are reasonably entitled to reject the vague contention of the armchair critic 'you can't prove it, there *may* be such a[n extenuating] feature [that results in a spurious correlation between smoking and lung cancer]'" was directed at Fisher's criticisms. Notice that an argument from ignorance is evident here. There is an inkling here of what will be discussed as pathological science in Chapter 4: Pathological Reasoning Within Sciences.

Rules-based and less formal WOE approaches are used in many cases but remain prone to error. They should be applied with the understanding that they produce weak inferences and that follow-up studies employing more powerful methods are required for more credible inference. The following quote from Hill's classic paper (Hill, 1965) is offered to any reader who disagrees with this last point.

Here then are nine different viewpoints from all of which we should study association before we cry causation. What I do not believe—and this has been suggested—is that we can usefully lay down some hard-and-fast rules of evidence that must be obeyed before we accept cause and effect. None of my nine viewpoints can bring indisputable evidence for or against the cause-and-effect hypothesis and none are required as sine qua non. What they do, with greater or less strength, is to help us to make up our minds on the fundamental question—is there any other way of explaining the set of facts before us, is there any other answer equally, or more, likely than cause and effect?

3.3 CONCLUSION

The goal of science is to organize knowledge around the most plausible explanatory principles. The central goal of the science of ecotoxicology is not to assess risk from a contaminant or to remediate a particular polluted site. Those are the technological and practical goals of environmental stewardship. Allowing ecotoxicology to efficiently pursue its scientific goal without being sidetracked will contribute to the laudable technical and practical goals in the long run.

There is no one valid scientific method although some methods are clearly better than others. However, there are accepted norms for what constitutes a scientific endeavor. As stated above, all scientific endeavors construct or assess explanations based on evidence, preferably from structured experiments, in a way that is open to scrutiny, challenge, or reinterpretation by others. Minimum subjectivity and maximum objectivity of practices and interpretations of evidence are essential features of scientific credibility. Practitioners reduce intersubjective bias by carefully applying the most powerful logical tools and then submitting their findings for external peer-review.

Many successful scientists conduct studies that involve only some aspects of the entire scientific process of inquiry. Some practice normal and others conduct innovative science. Together, they contribute to the same goal. In a young science like ecotoxicology, there is a tendency to emphasize normal science because a certain amount of sound evidence is needed before the innovative scientist can effectively judge existing and alternative explanations. Otherwise, weak testing with insufficient data could result in premature dismissal of sound explanations or hypotheses. As a young science transitions to a mature science, a preponderance of normal science is no longer ideal for its growth and the tyranny of the particular occurs for an awkward transition period: scientists remain fixated on gathering particulars and give insufficient attention to developing and testing robust explanations for the evidence. Scientists at this stage can also engage in excessive measurement, that is, fall victim to *idola quantitatus*. Skills associated with carefully formulating and testing new explanations are not emphasized, resulting in many weak inferences about hypotheses and explanations during this period. Ecotoxicology has been in such a transition period during which new explanations are not tested critically, resulting in weak inferences about hypotheses and explanations. Ecotoxicology does show signs of emerging from this stage in its evolution. Moderation of the tyranny of the particular and *idola quantitatus*, and more attention to innovative science skills are now warranted.

Another phenomenon influences the progress of ecotoxicology, that being, the Neurathian bootstrap. Quine (1960) describes the Neurathian bootstrap with an analogy, "[Otto] Neurath has likened science to a boat which, if we are to rebuild it, we must rebuild plank by plank while staying afloat in it ... Our boat stays afloat because at each alteration we keep the bulk of it intact as a going concern." Inferentially robust methods and explanations are essential in a healthy science but their application is constrained by the necessity of keeping the science mostly intact while slowly recrafting existing pieces. An extensive overhaul in a shipyard is a rarity only precipitated by a major crisis. In my opinion, ecotoxicology is overdue for a refitting because of a tradition of weak testing of many core paradigms and methods, and the Neurathian bootstrap (Newman, 2013). Also, concepts and methods facilitating regulatory ecotoxicology often are entangled with premises and methods necessary for the conduct of sound science.

The question asked about the SSD method at the onset can now be answered. None of the cited publications provides strong evidence for the ecological soundness of the SSD approach despite the widespread use of the approach for regulatory purposes. Its incongruity with fundamental ecological principles as pointed out by Forbes & Forbes (1993) and Hopkin (1993) have not been addressed rigorously. Also, the following fundamental issues from chapter 12 of Newman & Clements (2008) have not been resolved.

- LC50, EC50, NOEC and MATC data have significant deficiencies as measures of effect on natural populations or communities. Any method such as the SSD approach based on extrapolation from these metrics shares their shortcomings.
- The assumption that redundancy in communities permits a certain proportion of species to be lost is based on the redundant species hypothesis of synecology (Ehrlich & Walker, 1998). The alternate hypothesis, the rivet popper hypothesis, suggests the opposite would be true and that any loss of species weakens a community. Current evidence favors the rivet popper hypothesis. Until ecologists determine which is the most credible hypothesis, it would seem wise to be conservative and to adopt the rivet popper context for regulatory action.

- The SSD approach requires substantial knowledge of dominant and keystone species and the importance of species interactions. Adequate knowledge is often unavailable to the ecotoxicologist. Even when important species and species interactions are considered, consideration is often minimal.
- In situ exposure differs among species because of differences in their life histories, feeding habitats, life stages, microhabitat, and other qualities. These differences are poorly reflected in the species-sensitivity distribution method because of the toxicity test design. Most often toxicity tests are derived for one mode of exposure such as exposure from toxicant dissolved in waters.
- There is a bias toward mortality data despite the likelihood of sublethal effects playing a crucial role in local extinction of exposed populations.

Only one of the 22 chapters in the book, *Species Sensitivity Distributions in Ecotoxicology* (Posthuma, Suter & Traas, 2002) moved beyond embellishments of the approach (suggesting tyranny of the particular) and addressed the inferential strength of this method. The first sentence in the abstract of that single chapter (van den Brink, Brock & Posthuma, 2002) suggests a potential confirmation bias. "This chapter focuses on the field relevance of the output data of species sensitivity distribution (SSD) curves by seeking confirmation of laboratory-based SSD curves with population responses observed in semifield experiments." Like the weak support described earlier in studies by Emans et al. (1993), the explorations by van den Brink et al. (2002) provide weak support for the SSD approach.

The SSD approach has gained widespread use and credibility (Fig. 3.1) without repeated, strong testing of its fundamental soundness. This acceptance without thorough testing has the appearance of a suboptimal information cascade. Several plausible explanations for this deviation from norms of scientific inference are possible. There was an abundance of individual-based toxicity data available when the scientific and regulatory need emerged for a method to predict effects to ecological communities.[10] The Neurathian bootstrap came into play and the abundant toxicity test data were repurposed to infer effects to natural communities. Also, repeated survival of weak testing increased the credibility of the SSD approach, that is, familiarity bias fostered its acceptance. Inevitably, stronger tests will be done on the SSD approach, albeit after it has become entrenched, that might or might not support its utility. There are instances in autecology[11]

---

10. The consistency of a new idea with existing ones and ease of transition are qualities that foster adaptation as will be discussed from the innovation diffusion theory vantage in Chapter 7: How Innovations Enter and Move Within Groups.

11. Autecology is the study of the individual organism or species, and its relationships to its physical, chemical, and biological environment. Synecology involves studies of the interactions of groups of individuals. The study of population ecology is the boundary between autecology and synecology (Newman & Clements, 2008).

for which natural species abundances were adequately predicted from the physiological limits of individuals (Newman & Clements, 2008), and this might turn out to be the case for many SSD predictions. Or, reflecting many instances in synecology, the SSD approach might be deemed suboptimal as a general approach after stronger testing. Stronger testing of such widespread approaches, concepts, explanations, and theories needs to occur with increasing frequency as the science of ecotoxicology matures.

REFERENCES

Bacon, F. (1620). *Advancement of learning and Novum organum.* New York: Willey Book Company. (1944).

Beyers, D. W. (1998). Causal inference in environmental impact studies. *Journal of the North American Benthological Society, 17,* 367–373.

Biau, D. J., Jolles, B., & Porcher, R. (2010). P value and the theory of hypothesis testing. *Clinical Orthopedics and Related Research, 468,* 885–892.

Bollen, K. A., & Pearl, J. (2013). Chapter 15. Eight myths about causality and structural equation. In S. L. Morgan (Ed.), *Handbook of causal analysis for social research* (pp. 301–328). Dordrecht: Springer Science + Business Media.

van den Brink, P. J., Brock, T. C., & Posthuma, L. (2002). The value of the species sensitivity concept for predicting field effects: (Non-) confirmation of the concept using semifield experiments. In L. Posthuma, G. W. Suter, & T. P. Traas (Eds.), *Species sensitivity distributions in ecotoxicology* (pp. 155–193). Boca Raton: Lewis Publishers/CRC Press.

Carriger, J. F., & Newman, M. C. (2011). Influence diagrams as decision-making tools for pesticides. *Integrated Environmental Assessment and Management, 8,* 339–350.

Chamberlin, T. C. (1897). The method of multiple working hypotheses. *Journal of Geology, 5,* 837–848.

Ehrlich, P., & Walker, B. (1998). Rivets and redundancy. *Bioscience, 48,* 387.

Emans, H., Plassche, E., Canton, J., Okkerman, P., & Sparenburg, P. (1993). Validation of some extrapolation methods used for effect assessment. *Environmental Toxicology and Chemistry, 12,* 2139–2154.

Evans, A. S. (1976). Causation and disease: the Henle-Koch postulates revisited. *Yale Journal of Biology and Medicine, 49,* 175–195.

Feyerabend, P. (2010). *Against method (4th ed.).* New York: Verso.

Forbes, T. L., & Forbes, V. E. (1993). A critique of the use of distribution-based extrapolation models in ecotoxicology. *Functional Ecology, 7,* 249–254.

Fox, G. A. (1991). Practical causal inference for ecoepidemiologists. *Journal of Toxicology and Enviromental Health, 33,* 359–373.

Frampton, G. K., Jansch, S., Scott-Fordsmand, J. J., Rombke, J., & Van den Brink, P. J. (2006). Effects of pesticides on soil invertebrates in laboratory studies: a review and analysis using species sensitivity distributions. *Environmental Toxicology and Chemistry, 25,* 2480–2489.

Green, S. B., & Byar, D. P. (1984). Using observational data from registries to compare treatments: the fallacy of omnimetrics. *Statistics in Medicine, 3,* 361–370.

Hacking, I. (1996). The discontinuities of the sciences. In P. Galison, & D. J. Stump (Eds.), *The discontinuity of science* (pp. 37–74). Stanford: Stanford University Press.

Hill, A. B. (1965). The environment and disease: association or causation. *Proceedings of the Royal Society of Medicine, 58,* 295–300.

Hopkin, S. P. (1993). Ecological implications of '95% protection levels' for metals in soil. *OIKOS, 66*, 137−140.

Jesenska, S., Nemethova, S., & Blaha, L. (2013). Validation of the species sensitivity distribution in retrospective risk assessment of herbicides at the river basin scale—The Scheldt river basin case study. *Environmental Science and Pollution Research, 20*, 6070−6084.

Josephson, J. R., & Josephson, S. G. (1996). *Abductive inference. Computation, philosophy, technology*. Cambridge: Cambridge University Press.

Kooijman, S. A. (1987). A safety factor for LC50 values allowing for differences in sensitivity among species. *Water Research, 21*, 269−276.

Kuhn, T. S. (1962). *The structure of scientific revolutions (2nd ed.)*. Chicago: The University of Chicago.

Lane, D. A. (1989). Subjective probability and causality assessment. *Applied Stochastic Models and Data Analysis, 5*, 53−76.

Lane, D. A., Kramer, M. S., Huthinson, T. A., Jones, J. K., & Naranjo, C. (1987). The causality assessment of adverse drug reactions using a Bayesian approach. *Pharmaceutical Medicine, 2*, 265−283.

Last, J. M. (1983). *A dictionary of epidemiology*. Oxford: Oxford University Press.

Locke, J. (1690). *An essay concerning human understanding (collated by A.C. Fraser)*. New York: Dover Publications, Inc. (reprinted 1959).

Losee, J. (2001). *A historical introduction to the philosophy of science (4th ed.)*. Oxford: Oxford University Press.

Nagai, T., & Taya, K. (2015). Estimation of herbicide species sensitivity distribution using single-species algal toxicity data and information on the mode of action. *Environmental Toxicology and Chemistry, 34*, 677−684.

Newman, M. C. (1996). Ecotoxicology as a science. In M. C. Newman, & C. H. Jagoe (Eds.), *Ecotoxicology. A hierarchical treatment* (pp. 1−9). Boca Raton: CRC Press.

Newman, M. C. (2013). *Quantitative ecotoxicology (2nd ed.)*. Boca Raton: CRC Press/Taylor & Francis Group.

Newman, M. C., & Clements, W. H. (2008). *Ecotoxicology. A comprehensive treatment*. Boca Raton: CRC Press/Taylor & Francis Group.

Newman, M. C., Zhao, Y., & Carriger, J. F. (2007). Coastal and estuarine ecological risk assessment: the need for a more formal approach to stressor identification. *Hydrobiologia, 577*, 31−40.

Platt, J. R. (1964). Strong inference. *Science, 146*, 347−353.

Popper, K. R. (1959). *The logic of scientific discovery*. London: Routledge.

Posthuma, L., Suter, G. W., & Traas, T. P. (2002). *Species sensitivity distributions in ecotoxicology*. Boca Raton: Lewis Publishers/CRC Press.

Quine, W. V. (1960). *Word and object*. Cambridge: MIT Press.

Smetanova, S., Blaha, L., Liess, M., Schafer, R. B., & Beketov, M. A. (2014). Do predictions from species sensitivity distributions match with field data? *Environmental Pollution, 189*, 126−133.

Suter, G. W. (1993). A critique of ecosystem health concepts and indexes. *Environmental Toxicology and Chemistry, 12*, 1533−1539.

Suter, G. W. (2002). North American history of species sensitivity distributions. In L. Posthuma, G. W. Suter, & T. P. Traas (Eds.), *Species sensitivity distributions in ecotoxicology* (pp. 11−17). Boca Raton: Lewis Publishers.

Wilson, E. O. (1998). *Consilience. The unity of knowledge*. New York: John Wiley & Sons.

Chapter 4

Pathological Reasoning Within Sciences

Deviance, according to theories of social control, is not caused but is made possible because societal agencies can no longer prevent its occurrence.

Ben-Yehuda (1986)

4.1 PATHOLOGICAL SCIENCE

Chapter 3, Human Reasoning: Within Scientific Traditions and Rules, described the ways that science is best done, highlighting the steady progression to present-day norms for scientific inquiry. The scientific community has gradually evolved four core agencies that ensure members of a science do not deviate substantially from norms of good science: universalism, communality, disinterestedness, and organized skepticism (Merton, 1968). Universalism (objectivism) involves judging evidence or theory credibility without consideration of the person(s) reporting them. Our previous discussions exposed how difficult objectivity is to maintain at times for scientists. The second agency, communality, obligates scientists to make evidence and experimental details available for independent scrutiny by others. However, delays in publishing, demands of the moment, machinations to foster professional advancement, and basic social instincts can impede a scientist from quickly providing detailed information to others. The third agency, disinterestedness, requires that the scientist act for the sole purpose of extending scientific knowledge, not from other motives such as gaining financial profit, allowing a regulatory activity to proceed, or supporting personal core values such as environmentalism. The consequence to credibility if the scientist acts otherwise was expressed bluntly by Sarewitz (2012), "A biased scientific result is no different from a useless one." Organized skepticism, the fourth agency, involves open, constructive criticism and interpretation of evidence by others. Peer review and professional conferences are obvious venues for the expression of organized skepticism. Many factors can compromise constructive criticism. Acquiescence as described in Chapter 2, Human Reasoning: Everyday Heuristics and Foibles, can incline a potential critic to refrain from comment. At the other extreme, incivility or inattention to word

The Nature and Use of Ecotoxicological Evidence. DOI: https://doi.org/10.1016/B978-0-12-809642-0.00004-2

crafting can give criticism the outward appearance of a self-serving, personal attack. Careless expression of an otherwise insightful criticism can result in it being ignored by the scientific community and it also can provide a key ingredient for pathological sciences as described shortly.

Communality and organized skepticism, the second and fourth agencies, may require the ability to replicate studies. In practice, replication studies are only conducted to check particularly important claims, such as the claim that environmental concentrations of atrazine harm amphibians (Hayes et al., 2002; Rohr & McCoy, 2010; Van Der Kraak, Hosner, Hanson, Kloas, & Solomon, 2014), or extraordinary claims, such as the cold fusion claim by Fleischmann and Pons (1989). Except in these uncommon situations, there is little professional prestige to be gained by repeating the work of others. A grant application with the stated objective of replicating earlier work would have a very low likelihood of being funded.

These core agencies for controlling scientific behavior manifest through internal and external means (Ben-Yehuda, 1986). They are taught to and become internalized as central themes in a well-trained scientist's value system. External mechanisms include collective pressure in the scientific community to present one's work as objectively as possible, make it available for scrutiny by others, expect a shared skeptical vantage toward evidence or explanations, conduct research without personal motives, and be willing to accept the possibility of attempts to replicate one's scientific study. In everyday practice, habits and circumstances can hamper these internal and external control mechanisms, weakening their ability to prevent deviation from sound science. This short chapter explores the nature and different forms of deviant science.

To reliably identify deviations from acceptable scientific practice is difficult, so a few scenarios are bulleted below. Which do you think are (1) not science by intent nor approach, (2) science that went beyond the pale, (3) careless science, or (4) mistakes easily corrected as the science progresses?

- I published an article in which, among other results, bootstrap confidence intervals for species sensitivity distributions were interpreted incorrectly (Newman et al., 2000). A quick and civil exchange in the literature highlighted the error. Grist, Leung, Wheeler, and Crane (2002) described the shortcomings of the described bootstrap approach and provided a better one. From that point onward, anyone requesting detailed methods and the computer code for implementing the bootstrap intervals was redirected to the Grist et al. (2002) paper.
- Tyrone Hayes (2004) conducted a formal path analysis of the many conflicting research publications on the harmfulness of atrazine in the environment. He found that the likelihood of a study finding atrazine harmless depended mostly on whether the herbicide producer funded the study and also whether the study author had been a previous member of

an Ecorisk working group funded by the herbicide producer. Least important were study type (field or laboratory), study design rigor, and the
tested biological species. (A nonscientific treatment of the issue can be
found in a recent *The New Yorker* article (Aviv, 2014)). Dozens of
research papers have been published since, concluding that atrazine was
either harmful or harmless at environmental concentrations. A decade
after Hayes's (2004) publication, a group funded by the herbicide producer published a weight of evidence (WoE) review that countered assertions of harmfulness. The group concluded: "Overall, the WoE showed
that atrazine might affect biomarker-type responses ... at concentrations
sometimes found in the environment. However, these effects are not
translated to adverse outcomes in terms of apical [that is survival, growth,
development, and reproduction test] endpoints" (Van Der Kraak et al.,
2014). The declaration of interest in Van Der Kraak et al. (2014) states
that the authors were employees or consultants for the herbicide producer.
In some cases, authorship overlapped with membership of the original
Ecorisk working group.

- Understanding the importance of establishing the natural background
  concentrations and forms of metals in major rivers of the world, Ronald
  Gibbs (1973) measured Fe, Ni, Cr, Cu, and Mn associated with
  dissolved and particulate phases of the Amazon and Yukon Rivers.
  Dissolved metals were quantified with a chelation/extraction pre-
  concentration technique followed by an unspecified atomic absorption
  spectrophotometry (AAS) methodology. The extraction methods, clean
  technology, and AAS instrumentation were challenging at that time,
  and various flame and graphite furnace AAS techniques were being
  applied that differed in detection limits by orders of magnitude. Gibbs
  did not report detection limits, whether the data sets contained
  less-than-detection-limit (<DL) observations, nor how <DL observations might have been handled during computations of river metal loads.
  Although quantification was the main objective of the publication, accuracy (analytical bias) and precision were not reported.
- Burkholder, Noga, Hobbs, and Glasgow (1992)[1] reported a new dinoflagellate, later named *Pfiesteria piscicida* (Steidinger et al., 1996) that
  releases a potent neurotoxin after sensing the presence of fish and then
  feeds on the poisoned fish. They reported that it caused major estuarine
  fish kills along the North Carolina coast and could be the cause for the
  increasing number of unexplained fish kills occurring elsewhere.
  Anecdotal evidence was provided that exposure to the neurotoxin also
  affected human health. Publications attributing many North Carolina fish

---

1. See the book, *And the Waters Turned to Blood* (Barker, 1997), for a more detailed narrative
describing the associated studies by Burkholder, unpublished exchanges among local scientists
and regulators, and other factors related to this controversy.).

kills to low or depleted dissolved oxygen associated with eutrophication (Paerl, Pinckney, Fear, & Peierls, 1998) produced the following response, "[Paerl et al. (1998)] contains numerous misinterpretations and misuse of literature citations. Paerl *et al.* also make serious errors of omission, germane from the perspective of science ethics, in failing to cite peer-reviewed, published information that attributed other causality to various fish kills that they describe" (Burkholder, Mallin, & Glasgow, 1999). A series of published exchanges began, including claims of exaggerated human health risks (Griffith, 1999) and "falsely-based personal attacks" (Burkholder & Glasgow, 1999). The exchange continued in scientific publications like Vogelbein et al. (2002), Moeller et al. (2007), Patterson, Noga, and Germolec (2007), and Burkholder and Marshall (2012). Two decades after the first publication about this issue (Burkholder et al., 1992), the rancor is still apparent in Burkholder and Marshall (2012), "... beyond debate, there are major concerns that some publications continue to 'overlook', misinterpret, or misrepresent findings about *Pfiesteria* from other investigators."

- Morel's (1983) was the second most popular aquatic chemistry textbook with the first being Stumm and Morgan's (1970). Morel describes on page 42 of his textbook an equation dealing with mole fractions of a gas or solute, ending the description with an odd sentence, "For our purposes this equation is simply considered as God-given." If one looks up "God" in the book index, the following is provided, "God, 42, *See also* Morgan." Clearly, a playful jab at one of his professional rivals.

- During our earlier discussions of anchoring in Chapter 2, Human Reasoning: Everyday Heuristics and Foibles, recommendations of the Aquatic Dialogue group for pesticide risk assessment were mentioned (Aquatic Dialogue Group, 1994). The group funded by the US Environmental Protection Agency and the American Crop Protection Association, a not-for-profit trade organization composed of producers and distributors of pesticides and herbicides. The Group's intent when selecting the final tiered approach was clearly stated: "The process of [pesticide] risk assessment should be tiered to reduce the need for unnecessary assessments and delays in the regulatory process." The tacit assumption in their approach seems to have been that measures of toxicity to individual organisms can be applied to assure protection of ecological populations and communities. Protection of individuals was gauged expediently with available conventional toxicity testing data. Another assumption seems to have been that any ecological community will be protected if the predicted pesticide environmental concentration remains lower than that affecting 5% or 10% of the tested species.

- Baker et al. (1996) reported that two species of voles sampled near the Chernobyl site had mitochondrial cytochrome b base-pair substitution rates that were orders of magnitude higher than previously observed for

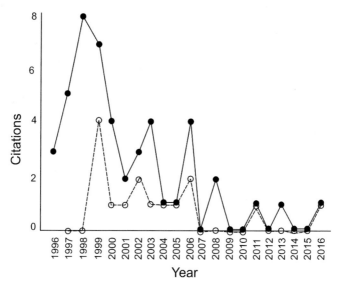

FIGURE 4.1   Citation frequency of the original report of exceptional rates of base-pair substitutions in mitochondrial cytochrome b (solid line) of voles and the retraction of the results published the next year (dashed line). Results are from a Web of Science search done on June 10, 2016; therefore, 2016 citation counts could be low. By mid-2016, there were a total of 47 and 14 citations to the original article and retraction, respectively. These results support several published findings that retractions do not efficiently correct the literature once mistakes are introduced (Budd, Sievert, Schultz & Scoville, 1999; Van Noorden, 2011). The increasing use of electronic forms of publications has improved this situation. With increasing frequency, journals are placing notifications of retraction on the electronic versions of retracted articles and providing the full citation for the retraction. No such notifications were made on the electronic version of this *Nature* article as of June 17, 2016.

any mammal. The numbers of individuals sampled for a species at the Chernobyl or the reference site were low (i.e., 4 or 5). The vole populations in the Chernobyl site seemed healthy despite this extraordinary base-pair substitution rate. Soon after publication, an error in manually sequencing the cloned DNA was discovered, leading to a paper retraction (Baker et al., 1997). Despite the published retraction, citation of the original paper continues to this date (Fig. 4.1).

4.1.1   Pathological Science Emerging From Human Foibles

Failure to faultlessly gather evidence and make judgments from that evidence is an inevitable feature of any human undertaking, including science. But, all scientific failings are not similar in how far they deviate from norms and impact the literature. They range from minor lapses, to unrecognized biases, full-blown pathological behavior, and blatant fraud.

Simple lapses are the most understandable type of error. Their impact varies greatly depending on how much a particular mistake influences the final evidence interpretation or how crucial the findings are in shaping the thinking about a particular topic or theory. They are not linked to any particular bias but how they are handled when recognized can be biased.

Reports of blatant fraud are scattered throughout the scientific disciplines, so they do not require much additional scrutiny here. However, two studies examining the general phenomenon of scientific fraud (and serious misbehavior) will be discussed briefly to provide the reader with a feeling for how frequently such behavior occurs. Martinson, Anderson, and de Vries (2005) surveyed 3247 scientists about their own behavior and that observed in others during the last 3 years. A very low number (3 of 1000) admitted that they themselves falsified or "cooked" research data.[2] Substantially more (60 of 1000) failed to present data that contradicted their own previous research. Even more (153 of 1000) removed data points based on a gut feeling of their inaccuracy. Based on these results, it would seem that falsifying evidence is universally envisioned as a major violation in science; yet, ignoring suspect or "bad" data is perhaps a forgivable lesser transgression. Fanelli (2009) conducted a meta-analysis of 21 surveys like that of Martinson et al. (2005). Fanelli found that 20 out of 1000 scientists self-reported having fabricated, falsified, or altered observations. Interestingly, the frequency was noticeably higher (141 of 1000) when scientists were asked whether they had observed such behavior by others. Perception of one's own and other's shortcomings was highly subjective.

In my experience (or perhaps, naivety), intentional fraud involving fabrication or falsification is rare in ecotoxicology and ecological risk assessment. Also, my opinion is that the cumulative effect of many instances of unintentional misrepresentation or misinterpretation of evidence is more harmful to ecotoxicology and ecological risk assessment than the occasional instance of outright fraud. Therefore, the emphasis here will be on unintentional errors that range from minor lapses to pathological behaviors.

As already discussed, subtle biases potentially influence scientific inquiry. Recollect that confirmation bias is a common tendency to seek confirmation of one's explanation and to discount disconfirming evidence. The 60 out of 1000 scientists who admitted in Martinson et al. (2005) to ignoring disconfirming data were victims of confirmation bias. Another bias, theory tenacity, involves the continued belief in a theory despite growing evidence of its failings (Loehle, 1987). Theory tenacity in science is fostered by the endowment effect, *status quo* bias, and anchoring error. Although theory tenacity is strictly speaking an inferential error, it can have a positive impact in a science by limiting the number of times that a scientific community swings from one theory

---

2. Data cooking is a term first applied nearly two centuries ago by Babbage (1830). He defined data cooking is the manipulation of data to make it appear to have better accuracy than true.)

to another. A healthy degree of resistance to change can reduce the time spent exploring spurious theories. A new piece of evidence might be flawed or initially misinterpreted in some way, so a certain degree of tenacity and scrutiny of contrary evidence are warranted before abandoning a theory. "A certain amount of dogmatism and pigheadedness is necessary in science if we are not to lose brilliant ideas which we do not at once know how to handle or to modify" (Popper, 1956). Another common error is enhanced belief in a theory because it has not been disproved during repeated weak testing, i.e., enhanced availability/credibility due to familiarity. In the worst case, this can lead to theory immunization in which credibility of a theory is inappropriately enhanced to such an extent that questioning it is seen as a fruitless endeavor. Difficulties in dealing with probabilities are yet another source of bias, including the pervasive misinterpretation of null hypothesis significance test in many sciences.

Another class of errors arises from shibboleths, that is, errors emerging from old or passé customs, beliefs, or positions held by subgroups of scientists. Bacon referred to them as idols of the theater, "various dogmas of peculiar systems of philosophy, ... [also included are the] many elements and axioms of sciences which have become inveterate by tradition, implicit credence, and neglect" (Bacon, 1620). He likens these dogmas to those adhered to by workers within civil governments where new ideas are avoided because "men must apply to them with some risk and injury to their own fortunes, and not only without reward, but subject to contumely and envy." This specific set of errors is pervasive in ecotoxicology and will be discussed below relative to the peculiar influence of regulatory activities on the conduct of science in our field.

To me, [it's] extremely interesting that men, perfectly honest, enthusiastic over their work, can so completely fool themselves. Now what is it about that work that made it so easy for them to do it?

Langmuir (1989)

Pathological science, the most serious unintentional deviation from acceptable scientific behavior, is the last error to be covered here. It is science practiced with insufficient objectivity. What is meant by "insufficient objectivity" is interpreted here as such a lack of objectivity that Merton's four agencies (mores or accepted norms of universalism/objectivity, communality, disinterestedness, and organized skepticism) can no longer constrain behavior to that expected of individuals in a scientific community. Unintentional subjectivity slowly grows to dominate evidence collection, analysis, and judging of explanation or theory plausibility. Langmuir's original description of pathological science (Langmuir, 1989) listed characteristics often present in cases of pathological science. Later, Rousseau (1992) added useful details. This list of pathological science qualities is intended to be used like Hill's rules of disease association. Beyond the absence of

TABLE 4.1 Eight Symptoms of Pathological Science

Six symptoms identified by Langmuir (1989) and Rousseau (1992)
1. A large effect is produced at barely detectable intensity of the causative agent with the effect being relatively independent of the causative agent intensity.
2. The important effect is near the limits of detection or the low statistical significance of the results requires large numbers of measurements. This point of Langmuir's can be generalized to the following: the notional effect is difficult to isolate from the signal noise or uncertainty, requiring many observations to produce a statistically significant effect.
3. There are claims of more precision (or accuracy) than plausible.
4. Prevailing ideas and theories are disregarded and fantastic theories contrary to experience suggested.
5. Criticisms or contrary evidence are countered with *ad hoc* explanations. Critical, definitive tests of the new explanation or theory are never done.
6. Support of the new explanation or theory increases to roughly 50% of the engaged scientific community and wanes thereafter.

Two symptoms not identified by Langmuir (1989) and Rousseau (1992)
7. A transition occurs in which the exchange of constructive criticism becomes, or is perceived to become, adversarial disparagements. Fortifying claims becomes the goal of evidence gathering and interpretation, not objectively deciding which is the most plausible of competing explanations or theories.
8. The legitimate goals and desirable qualities of other important activities are inappropriately substituted for or blended with those of science.

sufficient objectivity, no feature need be present to prove that pathological science exists. They are simply useful indicators that, if present, suggest that the associated science might be pathological. Langmuir and Rousseau's qualities are combined above into six points (Table 4.1). My impression after reviewing several documented instances of pathological science is that two other characteristics are often important. The first (adversarial dynamic) is a general one and the second (nonscientific context) is specifically associated with practical sciences such as ecotoxicology.

4.1.2 Sound Regulation Trumps Healthy Science?

Science is concerned with creating an intellectual model of the world. Technology is concerned with procedures and tools and their general use to gain or use knowledge. Practice is concerned with how to treat individual cases. Confusing the three can be dangerous.

Slabodkin and Dykhuizen (1991)

A distinctive type of error needs to be defined relative to the eighth quality of pathological science in Table 4.1. This context error emerges if the goals of science, technology, and practice become muddled in an applied science such as ecotoxicology (Newman, 1996, 2015).

To explain how a context error occurs, the goals of the science, technology, and practice of ecotoxicology need further clarification. The scientific goal of ecotoxicology is to generate, organize, and classify knowledge about pollutant fate and effects based on explanatory principles. Studies supporting this scientific goal adhere to the norms and methods already described. Related, but still distinct in important ways, are technological studies that have the goal of developing and then applying tools and methods for acquiring a better understanding of contaminant fate and effects in the biosphere. Qualities valued in technological studies are effectiveness (including cost-effectiveness), precision, accuracy, appropriate sensitivity, consistency, clarity of outcome, and ease of application (Newman, 2015). The third goal, the practical goal of ecotoxicology, is to apply available scientific knowledge and technologies to document or solve specific problems. In addition to the qualities valued in technological studies, unambiguous results, safety, and clear documentation of progress during application are also important features of practical ecotoxicology.

The usefulness of work in meeting one goal of ecotoxicology does not likewise make it useful for meeting that of another. A US EPA standard method for determining the concentration of a metal, say copper, in a water sample taken during National Pollutant Discharge Elimination System (NPDES) monitoring is an example of a technology that meets its intended goal in environmental regulation. The standard method has the desirable qualities for regulatory practice. However, its detection limit and inability to speciate copper forms might make it inadequate for a scientific study of copper geochemistry in an uncontaminated setting. Instead, a method that is not easy to apply might be used.

Cairns and co-investigators' assay for determining the impact of contaminants on habitat colonization dynamics by protozoans (Cairns, Pratt, Niederlehner, & McCormick, 1986; Niederlehner, Pratt, Buikema, & Cairns, 1985) represents the other extreme. Cairns' group developed a microcosm exposure system in which the rate of new habitat colonization was quantified with a MacArthur–Wilson island biogeography model. This microcosm approach had the undesirable quality (relative to a technology or practice) of requiring considerable taxonomic skill to identify aquatic protozoa and a sound ecological background in interpreting the findings. However, the assay provides a rare means of answering the scientific question of whether the contamination impacts ecological community processes.

A common context error in ecotoxicology involves the No and Lowest Observed Effect Concentration (NOEC and LOEC, respectively) metrics for regulatory purposes. The regulatory protocol for performing the related tests and computing these metrics (US EPA, 2002) incorporates the desirable qualities for a technology or practice: cost-effectiveness, defined accuracy and precision, consistency, and ease of application. However, this analysis of variance-based approach involving a wide range of toxicant concentrations is inferior to regression or change point analysis if one wishes to ask the science question, "Does a toxicant threshold concentration exist?" To answer that question, change point estimation (Chen & Gupta, 2012) might be applied to results from an experimental design with a narrow range of concentrations near the suspected threshold. Beyond this design limitation, the NOEC and LOEC metrics have minimal explanatory power: they provide little help in explaining observations of differences in species abundances between contaminated and uncontaminated ecosystems. To my knowledge, no series of rigorous, integrated studies has been completed that convincingly shows that NOEC and LOEC values relate to meaningful contaminant effects to ecosystems. Unfortunately, published studies purporting to be scientific investigations still apply the NOEC and LOEC metrics.

4.2 EXAMPLES OF PATHOLOGICAL SCIENCE

There are many documented instances of prominent scientists failing at times to practice the best science. A few will be discussed here to give the reader a flavor of such failings. The first will be that which Langmuir used in his initial description of pathological science. It will be explored in detail to illustrate how characteristics of pathological science can creep into scientific endeavors. After a few additional classic instances are described, the discussion will move away from comfortable, well-known examples of failings by—there is no nice way of saying this—scientists who are dead and can no longer take offense. Speculation about two apparent scientific failings in environmental sciences will be presented with the understanding that this speculation is an evidence-based opinion that might possibly be found false in the future. If this offends anyone, I remind them that the classic examples about to be described involved the father of modern genetics (Gregor Mendel), the father of modern statistics who was knighted for his work (Sir Ronald Fisher), and a Nobel laureate (Robert Millikan). Making one mistake does not permanently mark the associated scientist as somehow fatally flawed. Also, the point is not to criticize individuals but to examine factors that contribute to deviations from accepted norms.

4.2.1 Classic Examples

The clearest example is that used by Irving Langmuir (1989) to introduce the concept of pathological science. It involved a single experiment by a pair of scientists who later allowed Langmuir to examine their work as they replicated a critical experiment. The details from the original research seminar describing the experiment and later laboratory visit reveal how well-intended scientists fall prey to pathological behavior. As characteristics of pathological science from Table 4.1 emerge in this summary of Langmuir's narrative, the number of each will be placed in parentheses, such as (Characteristic 2).

Langmuir related a story which began when a seminar speaker, Bergen Davis, presented his novel findings with co-investigator, Arthur Barnes, at the General Electric Laboratory in Schenectady, New York. Dissatisfaction with post-seminar explanations of several aspects of their findings prompted Langmuir to request that he be allowed to observe as Davis demonstrated the experimental system. Davis was a faculty member at nearby Columbia University, so a visit was easily arranged.

Before delving into the physical details of the experimental system, an important point needs to be made that was not apparent in Langmuir's account. Langmuir related that Davis was "glad" and "proud" to provide a laboratory demonstration; however, it is hard to imagine that Davis was not also very anxious. The laboratory visit would take place in 1930, just 2 years before Langmuir was awarded a Nobel Prize in chemistry. Langmuir was a very formidable figure at the time. Langmuir's description of the visit reflects only collegiality and civility with no mention of defensiveness on Davis's part when quizzed by this intimidating scientist. Also, the visit was prompted by Willis Whitney, the General Electric laboratory director, who was interested in the experiment being repeated after replacing the zinc sulfide screen with a more accurate Geiger counter that Clarence Hewlett had in his General Electric laboratory. Langmuir suggested that he and Hewlett visit Davis's laboratory for a demonstration before the Geiger counter was made available. The possibility of gaining access to better equipment surely added to Davis's anxiety about the visit. In my opinion, it is extremely important to keep in mind as one proceeds through the example that Langmuir seemed oblivious to this key aspect of Davis' emotional state during the seminar and site visit (Characteristic 7).

Details of Davis and Barnes' research setup are relatively straightforward to describe (Fig. 4.2). The experiment involved α-particles (i.e., helium nuclei with a $+2$ charge) streaming through a hole in the center of a parabolic hot cathode electron emitter (filament). The filament emitted electrons that were accelerated by the anode grid and focused toward a downstream anode. The electron density in the resulting stream was modulated by current density. Davis and Barnes adjusted the potential to 590 V, so that the resulting stream of emitted electrons would have the same velocity as the doubly charged α-particles passing through the cathode hole. They hoped that some electrons and α-particles in the coincident beams would combine, changing the charge of some particles to $+1$. To ascertain the number of uncombined and combined particles, the particle beam passed through a magnetic field that deflected them to a degree determined by their charge (and mass). The number of particles deflected to different degrees was estimated by counting faint flashes as particles hit a zinc sulfide screen placed in the beam path. A skilled observer attempting such a count was required to accommodate their eyes for half an hour in a completely darkened room. During counting, he also had to recognize and dismiss flashes resulting from the decay of natural radioactive

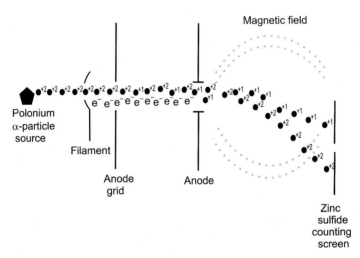

FIGURE 4.2   Schematic of the apparatus used by Davis and Barnes.

contaminants in the system (Characteristic 2). If all went well, any change in particle energy after acquiring an electron was to be compared to expectations from the Rutherford—Bohr theory of the hydrogen atom.

Davis explained in his seminar that the change in number of flashes when a 590-V potential was applied to the cathode indicated that approximately 80% of the α-particles combined with electrons. This 80% was relative to the number counted when there were no electrons or magnetic field generated during a reference trial. Surprisingly, 80% also combined with electrons when the cathode was set at other discrete voltages. These discrete voltages corresponded exactly to those at which an electron velocity was equal to that which it would have in a Bohr orbital.

Langmuir had difficulty with these results. An electron must give up energy to move from outside the atom into an orbital, but the results suggested no such energy loss (Characteristic 4). Also, the percentage of particles combining with electrons was independent of the current density, that is, of the density of electrons in the coincident beams available to combine with α-particles (Characteristic 1). During the questioning, Langmuir pointed out to Davis that his claims were being made even for apparatus settings at which no electrons would be emitted from the filament. Davis quickly responded that there would still be sufficient numbers of electrons under the conditions being questioned (Characteristic 5). The percentage of combined particles was independent of the electron density, including densities approaching zero! Questions also emerged about how they had discovered the precise voltages at which peak electron capture was occurring (Characteristic 3). The time needed to slowly count flashes at very narrow intervals of 1/100 V in the relevant spectrum ranging from 300 to 1000 V brought questions about how the laborious peak identifications were done. Also, the precision of the voltages that exactly matched Bohr's predictions seemed beyond the capabilities of the employed equipment and the field produced by the apparatus would likely have been insufficiently uniform to support such claims of precision (Characteristic 3). Throughout the questioning, Davis provided *ad hoc* answers to queries of this nature that, on later reflection, seemed inadequate (Characteristic 5).[3]

Their claim that α-particles and electrons came together at energies predicted by the Rutherford–Bohr theory completely unraveled during the site visit. Langmuir and Hewlett sat alongside Davis in a darkened room as their eyes adjusted before practicing flash counting. Langmuir and Hewlett got fairly consistent counts with and without a potential being applied. The background counts when the potential was applied to deflect particles away from the screen were quite high, that is, 25% of the count when no potential was applied. Relative to counts by Langmuir and Hewlett, Barnes detected many more flashes with no applied potential and similar counts when a potential was applied. The visitors were told by Barnes that he counted flashes for 2 minutes, but Langmuir used a stopwatch in the darkened room to discover that Davis' count times actually ranged from 70 to 115 seconds (Characteristic 3). During the trails and counts, the voltages applied to the anode were adjusted by an assistant to within a $1/100^{th}$ V on a voltmeter with a scale from 1 to 1000 V (Characteristic 3).

In his own words, Langmuir then "played a dirty trick" on Barnes. He produced written cards directing an assistant to either apply or not apply voltage prior to Barnes beginning his count of flashes. He randomly passed a card to the assistant who did as instructed before each count. As the test continued, it was clear that the assistant's movements provided hints that Barnes subconsciously noted as he began counting. The assistant sat back in his chair when the card indicated that he was to leave to voltage off but sat forward to adjust the voltage to a specific value when instructed to apply voltage. There is no indication in Langmuir's narrative that Barnes was consciously influenced by these cues. But, as described in detail by Mlodinow (2012), subconscious reading of others is a pervasive feature in human interactions and not to be discounted. Any difference in counts vanished when the assistant was instructed by Langmuir to act identically in all situations. Barnes responded with an ad hoc excuse for this particular instance of failure. He decided that the vacuum tube containing the apparatus had too much gas in it to function properly (Characteristic 5).

---

3. Viewed from the context of a formal scientific exchange, it might be difficult to understand why others listening to the seminar did not object to Barnes's ad hoc responses. However, it is easily understood in terms of the acquiescence bias described in Chapter 2, Human Reasoning: Everyday Heuristics and Foibles. People tend to accept something as it is presented. They have an instinctual desire to be collegial and civil, and to avoid confrontation.

It also became obvious during discussions that they had calculated the position of potential peaks using Bohr theory before searching for them with the apparatus to "save time" in searching through the entire spectrum. Counting had subconsciously been influenced by this knowledge.

Neither, Barnes nor Davis was convinced that they were wrong and continued to present their findings after the visit. Langmuir sent a letter to Niels Bohr, asking that it be passed on to others who might be considering spending time further exploring what Langmuir now believed was unsound science. A year later, the issue was resolved when Davis and Barnes published a brief article stating that they were unable to reproduce their original findings.

It was important to review this example because it highlights qualities of pathological science nicely. The fine detail allowing a definitive conclusion results from the direct involvement of the narrator, Langmuir, in the process and the openness of the researchers to scrutiny. Less definitive examples are still useful in identifying how unintentional subjectivity can confuse scientific inquiry, even though all qualities of pathological science might not be pinpointed. Three examples are provided to illustrate different degrees of deviation from acceptable scientific norms. It is left up to the reader to decide which instances reflect genuine pathological science. The findings of two of the studies are consistent with those of follow-up studies. One, Fisher's arguments about smoking and lung cancer, has clear symptoms of pathological science in my opinion.

With uncharacteristic tact, Ronald Fisher (1936) demonstrated that Gregor Mendel's pea genetics data were improbably close to binomial expectations. This began the Mendel–Fisher controversy that continues to this day (Radick, 2015). The speculated root cause of these too-perfect data range from Mendel's assistant innocently cooking the data to Mendel's subconscious bias in favor of binomial expectations (Dworkin, 1983). Whether pathological science was present here remains a matter of conjecture because those directly involved are no longer alive to be questioned. To be completely fair to Mendel, improbable events such as his data set do occur.

Robert Millikan won a Nobel Prize in 1923 for estimating the discrete charge of electrons. In his published research employing oil droplets suspended in a magnetic field, he stated that all droplets were used; however, posthumous examination of his notebooks showed that data were discarded for 49 "bad" droplets of a total of 140 droplets (Ayala et al., 1989; Goodstein, 1992). Although his conclusions and calculated electron charge are consistent with later studies by others, the claim of using all observations and lack of clarity about what constituted a "bad" observation have stimulated many discussions on scientific ethics. It remains a point of speculation whether this was an instance of pathological science or simply poor documentation of droplet selection criteria.

With characteristic lack of tact, Ronald Fisher (1957, 1958a,b,c) objected to early epidemiological studies linking smoking to lung cancer. Fisher, a long-time smoker and occasional consultant to the tobacco industry (Stolley, 1991), used his statistical genius to emphasize the inadequacies of WoE approaches for drawing definitive inferences about disease causality. His logic was correct as far as it went; however, he failed to acknowledge the value of such rules in identifying plausible explanations deserving more rigorous study. He proposed an alternative genetic explanation that deflected inquiry away from the original question for a time. He hypothesized that the correlation between lung cancer and smoking resulted from an unknown genetic factor that predisposed some individuals to smoking and lung cancer. Perhaps, his fondness for pipe smoking and association with the tobacco industry created a value centrality bias that blinded him to the positive value of qualitative WoE methods. Regardless, his incorrect arguments were a blend of elegant statistical logic, premature conclusions, and selective use of available evidence.

4.2.2 Environmental Science Examples

Enough explanation of pathological science has been given at this point to make some judgments about the examples provided in the beginning of this chapter. The easiest example to dismiss is the obtuse jab by Morel, referring to his competitor as "god." It is very difficult to see how this jab would reduce the effectiveness of the four agencies identified by Merton (1968) as ensuring good science. The examples of Newman et al. (2000) and Baker et al. (1996) seem instances of straightforward, albeit careless, errors that were corrected by the scientific process. The lack of specific details makes it difficult to judge the omission of analytical details by Gibbs (1973). However, like the Millikan example just described, the paucity of detail makes it difficult for the Mertonian communality agency to come into play. It would be difficult to independently evaluate the results in the absence of the important analytical details like detection limits and how $<$DL observations were handled during calculations. The remaining two examples are less easily dismissed and will be discussed in more detail. Unlike the Langmuir example in which one of the parties (Langmuir) was an apparently impartial judge, it is difficult to pinpoint clear pathological scientists in these examples because impartiality becomes difficult to attribute to any single actor in the exchanges. Instead of identifying erring individuals, evidence suggesting loss of adequate objectivity will be emphasized in the discussions below.

Evidence of substantial deviation from norms of objective science exists in the published attempts to identify the cause for large fish kills in North Carolina. The book, *And the Waters Turned Red*, written about the scientists and state resource managers involved in early stages of the process (Barker, 1997) contains numerous descriptions of subjectivity. Belousek (2004) also describes the insertion of regional politics into the scientific process of finding the reason for coastal fish kills. He observes that the blending of scientific and political agencies resulted in "an anti-correlation between the universality of the mandated scientific consensus and the political relevance of the consensus." The publication by Paerl et al. (1998) proposing eutrophication-driven low dissolved oxygen concentrations as an alternative explanation for fish kills prompted the claim that they made "serious errors of omission, germane from the perspective of science ethics ..." (Burkholder et al., 1999). David Griffith (1999) wrote, "Despite mounting evidence that *Pfiesteria piscicida*, a marine organism that releases a neurotoxin, poses no threat to public health, its threat continues to be exaggerated by journalists, popular writers, politicians, and scientists." Burkholder and Glasgow (1999) responded that the Griffith article contained "falsely-based personal attacks." Belousek (2004) quotes Joseph Ramus of Duke University as concluding that the entire process amounted to "the total failure of a principle of science. The peer-review process has been circumvented ... it has collapsed."

My own opinion, after reading 48 journal articles, a patent for the *Pfiesteria* toxin, and Baker's popular book about the notionally *Pfiesteria*-linked fish kills, is that pathological science exists in this instance. The published literature seems more of a long sequence of point and counterpoint exchanges than objective, evidence-based testing by the scientific community of plausible explanations until only the most plausible explanation(s) remained. Relating back to the previous list of pathological science characteristics, there was evidence of Characteristics 4 (prevailing theory ignored) and 5 (criticisms and contrary evidence met with *ad hoc* explanations). More than 15 years after publication of the first article proposing *Pfiesteria* as a cause for many large fish kills, there remain many supporters and detractors of this hypothesis. There seem to be fewer supporters as time goes by (Characteristic 6, increase in support of new theory and then gradual decline). But most obvious are a context error that compromised the scientific validity of the endeavor (Characteristic 8) and the transition from a civil process of constructive criticism to an adversarial exchange among scientists (Characteristic 7).

> *Science is a principle and a process of seeking truth. Truth cannot be purchased and, thus, truth cannot be altered by money.*
>
> Hayes (quoted in Aviv, 2014)

> *Mark Schlissel, dean of biological sciences at the University of California, Berkeley, says that the "decade long" dispute between Hayes and Syngenta has "moved increasingly into the personal arena".*
>
> Dalton (2010)

In my opinion, the environmental atrazine example is another instance of pathological science, that is, science practiced in the absence of adequate objectivity. The formal path analysis done by Hayes (2004) addressed the question of impartiality in atrazine studies, concluding that the likelihood of a study finding atrazine harmless depended most heavily on whether the research being published was funded by the producer and whether the investigator was a past member of a producer-funded Ecorisk working group.

Numerous publications have demonstrated the potential for harm from atrazine including a thorough meta-analysis of atrazine's effects to fish and amphibians (Rohr & McCoy, 2010). A recent publication (Vandenberg et al., 2013) makes the reasonable argument that regulatory decisions about endocrine system modifiers such as atrazine should be based primarily on endocrinologic evidence. Although ample evidence exists about impacts of environmental levels of atrazine on the endocrine system (Rohr & McCoy, 2010), a recent Syngenta-funded WoE study (Van Der Kraak et al., 2014) concluded that atrazine's effects "were not translated to adverse outcomes in terms of [survival, growth, development or reproduction]." In contrast to the United States which still allows atrazine application, the European Union halted its use (Hakim, 2015) based on existing evidence.

These contrasting evidences from different sources prompted Tyrone Hayes to expand his role from scientist to advocate. The Tyrone Hayes and Penelope Jagessar Chaffer's December 2010 "Toxic Baby" TED talk at https://www.ted.com/speakers/tyrone_hayes makes the reasons for this transition clear. Motivation for the transition can also be found in Aviv (2014) or Hayes (2004). Hayes was reported to have sent offensive e-mails to Syngenta employees, leading Syngenta to appeal directly to Hayes' home university (University of California-Berkeley) about e-mail ethical standards. The Syngenta website was cited by Dalton (2010) as referring to Hayes's work as "questionable studies." Clearly, the process of a scientific community collectively and objectively studying the effects of environmental concentrations of atrazine has been, and still is, subverted by context biases. The individuals who depend on efficient scientific generation and application of sound information about pollutants in the biosphere were, and still are, poorly served in the process.

4.3 CONCLUSION

The agencies that Merton (1968) outlined seem reasonably effective at assuring that most science is practiced in a healthy manner. Universalism (objectivity) should minimize subjectivity especially if Chamberlin's multiple working hypothesis approach is conscientiously practiced. Scientists should report their studies with sufficient detail and clarity so that others may carefully scrutinize the results. The resolution of Langmuir's first example of pathological science would have not been possible without the willingness of Davis and Barnes to open themselves up to scrutiny by Langmuir and Hewlett. Communal skepticism requires that a scientist participate in the process of constructive criticism in a civil manner. I was personally struck by the contrast between the civility of the exchange reported by Langmuir and the incivility evident in the atrazine harmfulness and *Pfiesteria*-linked fish kill examples. The agency of disinterestedness was clearly missing from these latter examples. The likelihood of pathological science is minimized if young ecotoxicologists are taught and older ecotoxicologists practice these four agencies as essential features of sound science, not simply ivory tower niceties.

4.3.1 Pathological Science in Our Applied Science

Several themes emerge in the examples of pathological science in ecotoxicology. The first and most important is the high risk of a context bias. Taking on a non-science context such as those otherwise worthy contexts associated with government regulation, agrichemical product licensing, or environmental advocacy is asking for trouble if one purports to be doing science. No misdirected argument that these contexts are valid will change the fact that they are not the proper context for healthy scientific inquiry. The risk of context bias is extremely high in ecotoxicology.

An expectation of constructive criticism is central to a healthy science. In my opinion, openness to criticism is mandatory although extremely difficult to accept about one's own work. It is unfortunate that a minority of anonymous reviewers of manuscripts and proposals ignore this and fall victim to the social elimination error.[4] The reviewer makes the flawed judgment that a shortcoming or error in a study or proposal is evidence that the researcher is flawed and unworthy of civility. Unfortunately, several scholars have noted an increase in incivility among academics and a turn toward more unscientific tactics. Books documenting the increase in academic incivility and possible ways for lessening its impact include Hollis (2012), Twale and De Luca (2008), and Westhues (1998; 2004; 2006).

> *Organizational survival-of-the-fittest norms often encourage micropolitics through the use of uncivil, bully, and mob behaviors. In these instances, discussion turns to argumentation, posturing, and contention.*
>
> Twale and De Luca (2008)

This general trend in academia, where much of science is conducted, tempers any hope that incivility in scientific discourse will be remedied soon.

---

4. Westhues (2004) identifies the phenomena of social elimination as being a fundamental attribution error (also called correspondence error) in which an individual or group "overestimate[s] the extent to which behavior reflects underlying personal qualities (dishonesty, for instance, or untrustworthiness) and to underestimate the extent to which it reflects the particular situation or context." The scientific error or shortcoming is erroneously translated to the researcher's unworthiness. This cognitive error was not described in Chapter 2, Human Reasoning: Everyday Heuristics and Foibles, because it deals with processing social evidence about individuals or groups, not physical evidence.

4.3.2 Minimizing Pathological Ecotoxicology

Pathological science can emerge in ecotoxicology for the same reasons it might in other sciences. Individual or teams of scientists can slowly move from objective inquiry to subjective studies attempting to support their favored conjecture. Ecotoxicology has emerged as a science during a period of increasing academic incivility and intense competition for limited funding. This incivility and intense competition can increase the risk of falling away from objective science. Also extremely influential in an applied science like ecotoxicology is context error. The laudable context of some activities such as environmental regulation or advocacy can result in dysfunctional scientific activities if inserted carelessly into evidence generation and interpretation.

The risk of pathological science can be reduced in several ways. Foremost is the teaching of and adherence to the rules of healthy scientific endeavors (i.e., the Mertonian agencies of universalism, communality, disinterestedness, and organized skepticism). The goal of and tenants for conducting healthy science should not be confused with those of other activities. Being swayed in the conduct of science by well-intended arguments (e.g., scientists must also be advocates) can jeopardize the credibility of one's science in the long run, so it is critical to know when one has moved away from doing science to another activity. Lastly and very importantly, civility is essential to avoid deviance from the norms of science.

REFERENCES

Aviv, R. (2014). *A valuable reputation. After Tyrone Hayes said that a chemical was harmful, its maker pursued him* (pp. 1–18). The New Yorker.

Ayala, F., Adams, R. M., Chilton, M.-D., Hull, D., Patel, K., Press, F., ... Sharp, P. (1989). *On being a scientist*. Washington, DC: National Academy of Science.

Babbage, C. (1830). Reflections on the decline of science in England and on some of its causes. In M. Campbell-Kelly (Ed.), *The works of Charles Babbage*. London: Pickering.

Bacon, F. (1620). *Advancement of learning and Novum organum*. New York (1944): Wiley Book Company.

Baker, J. L. (1994). *Aquatic dialogue group: Pesticide risk assessment & mitigation*. Pensacola: SETAC Press.

Baker, R. J., Van Den Bussche, R. A., Wright, A. J., Wiggins, L. E., Hamilton, M. J., Reat, E. P., ... Chesser, R. K. (1996). High levels of genetic change in rodent of Chernobyl. *Nature, 380*, 707–708.

Baker, R. J., Van Den Bussche, R. A., Wright, A. J., Wiggins, L. E., Hamilton, M. J., Reat, E. P., ... Chesser, R. K. (1997). Retraction. High levels of genetic change in rodents of Chernobyl. *Nature, 390*, 100.

Barker, R. (1997). *And the waters turned to blood*. New York: Simon & Schuster Paperbacks.

Belousek, D. W. (2004). Scientific consensus and public policy: The case of *Pfiesteria*. *The Journal of Philosophy, Science & Law, 4*, 1–33.

Ben-Yehuda, N. (1986). Deviance in science. *The British Journal of Criminology, 26*, 1–17.

Budd, J. M., Sievert, M., Schultz, T. R., & Scoville, C. (1999). Effects of article retraction on citation and practice in medicine. *Bulletin of the Medical Library Association, 87*, 437–443.

Burkholder, J. M., & Glasgow, H. B. (1999). Science ethics and its role in early suppression of the *Pfiesteria* issue. *Human Organization, 58,* 443–455.

Burkholder, J. M., Mallin, M. A., & Glasgow, H. B. (1999). Fish kills, bottom-water hypoxia, and the toxic *Pfiesteria* complex in the Neuse River and estuary. *Marine Ecology Progress Series, 179,* 301–310.

Burkholder, J. M., & Marshall, H. G. (2012). Toxigenic *Pfiesteria* species—Updates on biology, toxins, and impacts. *Harmful Algae, 14,* 196–230.

Burkholder, J. M., Noga, E. J., Hobbs, C. H., & Glasgow, H. B. (1992). New 'phantom' dinoflagellate is the causative agent of major estuarine fish kills. *Nature, 358,* 407–410.

Cairns, J. J., Pratt, J. R., Niederlehner, B. R., & McCormick, P. V. (1986). A simple cost-effective multispecies toxicity test using organisms with a cosmopolitan distribution. *Environmental Monitoring and Assessment, 6,* 207–220.

Chen, J., & Gupta, A. K. (2012). *Parametric statistical change point analysis* (2nd ed.). New York: Springer Science + Business Media, LLC.

Dalton, R. (2010). E-mails spark ethics row. *Nature, 466,* 913.

Dworkin, G. (1983). Fraud and science. *Progress in Clinical and Biological Research, 128,* 65–74.

Fanelli, D. (2009). How many scientists fabricate and falsify research? A systematic review and meta-analysis of survey data. *PLoS One, 4,* e5738.

Fisher, R. A. (1936). Has Mendel's work been rediscovered? *Annals of Science, 1,* 115–137.

Fisher, R. A. (1957). Dangers of cigarette-smoking. *British Medical Journal,* 297–298, Aug 3.

Fisher, R. A. (1958a). Cigarettes, cancer, and statistics. *Centennial Review, 2,* 151–166.

Fisher, R. A. (1958b). Lung cancer and cigarettes? *Nature, 182,* 108.

Fisher, R. A. (1958c). The nature of probability. *Centennial Review, 2,* 261–274.

Fleischmann, M., & Pons, S. (1989). Electrochemically induced nuclear fusion of deuterium. *Journal of Electroanalytical Chemistry, 261,* 301–308.

Gibbs, R. J. (1973). Mechanisms of trace metal transport in rivers. *Science, 180,* 71–73.

Goodstein, D. (1992). What do we mean when we use the term 'science fraud'? *Scientist, 3,* 11–12.

Griffith, D. (1999). Exaggerating environmental health risk: The case of the toxic dinoflagellate *Pfiesteria. Human Organization, 58,* 119–127.

Grist, E. P., Leung, K. M., Wheeler, J. R., & Crane, M. (2002). Better bootstrap estimation of hazardous concentration thresholds for aquatic assemblages. *Environmental Toxicology and Chemistry, 21,* 1515–1524.

Hakim, D. (2015). A pesticide banned, or not, underscores trans-Atlantic trade sensitivities. *The New York Times.* Available from http://nti.ms/1A22MTC.

Hayes, T., Haston, K., Tsui, M., Hoang, A., Haeffele, C., & Vonk, A. (2002). Feminization of male frogs in the wild. *Nature, 419,* 895–896.

Hayes, T. B. (2004). There is no denying this: defusing the confusion about atrazine. *BioScience, 54,* 1138–1149.

Hollis, L. P. (2012). *Bully in the ivory tower.* St. Louis, MO: PatricaBerkly LLC.

Langmuir, I. (1989). *Pathological science* (pp. 36–48). Physics Today.

Loehle, C. (1987). Hypothesis testing in ecology: Psychological aspects and the importance of theory maturation. *The Quarterly Review of Biology, 62,* 397–409.

Martinson, B. C., Anderson, M. S., & de Vries, R. (2005). Scientists behaving badly. *Nature, 435,* 737–738.

Merton, R. K. (1968). *Social theory and social structure.* New York: The Free Press.

Mlodinow, L. (2012). *Subliminal.* New York: Pantheon Books.

Moeller, P. D., Beauchesne, K. R., Huncik, K. M., Davis, W. C., Christopher, S. J., Riggs-Gelasco, P., & Gelasco, A. K. (2007). Metal complexes and free radical toxins produced by *Pfiesteria piscicida*. *Environmental Science & Technology*, *41*, 1166−1172.

Morel, F. M. (1983). *Principles of aquatic chemistry*. New York: John Wiley & Sons.

Newman, M. C. (1996). Ecotoxicology as a science. In M. C. Newman, & C. H. Jagoe (Eds.), *Ecotoxicology. A hierarchical treatment* (pp. 1−9). Boca Raton: CRC Press.

Newman, M. C. (2015). *Fundamentals of ecotoxicology. The science of pollution* (*4th Edition*). Boca Raton: CRC Press/Taylor & Francis Group.

Newman, M. C., Ownby, D. R., Mezin, L. C., Christensen, T. R., Lerberg, S. B., & Anderson, B.-A. (2000). Applying species-sensitivity distributions in ecological risk assessment: Assumptions of distribution type and sufficient numbers of species. *Environmental Toxicology and Chemistry*, *19*, 508−515.

Niederlehner, B. R., Pratt, J. R., Buikema, A. L., & Cairns, J. (1985). Laboratory tests evaluating the effects of cadmium on freshwater protozoan communities. *Environmental Toxicology and Chemistry*, *4*, 155−165.

Paerl, H. W., Pinckney, J. L., Fear, J. M., & Peierls, B. L. (1998). Ecosystem responses to internal and watershed organic matter loading: consequences for hypoxia in the eutrophying Neuse River estuary, North Carolina, USA. *Marine Ecology Progress Series*, *167*, 17−25.

Patterson, R. M., Noga, E., & Germolec, D. (2007). Lack of evidence for contact sensitization by *Pfiesteria* extract. *Environmental Health Perspectives*, *115*, 1023−1028.

Popper, K.R. (1956). *Realism and the aim of science*. (Trans. 1983). New York: Routledge.

Radick, G. (2015). Beyond the "Mendel-Fisher controversy". *Science*, *350*, 159−160.

Rohr, J. R., & McCoy, K. A. (2010). A qualitative meta-analysis reveals consistent effects of atrazine on freshwater fish and amphibians. *Environmental Health Perspectives*, *118*, 20−32.

Rousseau, D. L. (1992). Case studies in pathological science. *American Scientist*, *80*, 54−63.

Sarewitz, D. (2012). Beware the creeping cracks of bias. *Nature*, *485*, 149.

Slobodkin, L. B., & Dykhuizen, D. E. (1991). Applied ecology: its practice and philosophy. In J. J. Cairns, & T. V. Crawford (Eds.), *Integrated environmental management* (pp. 63−70). Chelsea: Lewis Publishers.

Steidinger, K. A., Burkholder, J. M., Glasgow, H. B., Hobbs, C. W., Garrett, J. K., Truby, E. W., ... Smith, S. A. (1996). *Pfiesteria piscicida* GEN. ET SP. NOV (Pfiesteriaceae FAM. Nov.) A new toxic dinoflagellate with a complex life cycle and behavior. *Journal of Phycology*, *32*, 157−164.

Stolley, P. D. (1991). When genius errs: R.A. Fisher and the lung cancer controversy. *American Journal of Epidemiology*, *133*, 416−425.

Stumm, W., & Morgan, J. J. (1970). *Aquatic chemistry. An introduction emphasizing chemical equilibria in natural waters*. New York: John Wiley & Sons.

Twale, D. J., & De Luca, B. M. (2008). *Faculty incivility. The rise of the academic bully culture and what to do about it*. Hoboken, NJ: John Wiley & Sons, Inc.

US Environmental Protection Agency. (2002). *Short-term methods for estimating the chronic toxicity of effluents and receiving waters to freshwater organisms*. Washington: US Environmental Protection Agency, EPA-821-R-02-013.

Vandenberg, L. N., Colborn, T., Hayes, T. B., Heindel, J. J., Jacobs, D. R., Lee, D.-H., et al. (2013). Regulatory decisions on endocrine disrupting chemicals should be based on the principles of endocrinology. *Reproductive Toxicology*, *38*, 1−15.

Van Der Kraak, G. J., Hosner, A. J., Hanson, M. L., Kloas, W., & Solomon, K. R. (2014). Effects of atrazine in fish, amphibians, and reptiles: An analysis based on quantitative weight of evidence. *Critical Reviews in Toxicology, 44*, 1–66.

Van Noorden, R. (2011). The trouble with retractions. *Nature, 478*, 26–28.

Vogelbein, W. K., Lovko, V. J., Shields, J. D., Reece, K. S., Mason, P. L., Haas, L. W., & Walker, C. C. (2002). *Pfiesteria shumwayae* kills fish by micropredation not exotoxin secretion. *Nature, 418*, 967–970.

Westhues, K. (1998). *Eliminating professors. A guide to the dismissal process.* Queenston, Ontario: The Edwin Mellen Press.

Westhues, K. (2004). *Workplace mobbing in academe. Reports from twenty universities.* Queenston, Ontario: The Edwin Mellen Press.

Westhues, K. (2006). *Remedy and prevention of mobbing in higher education.* Queenston, Ontario: The Edwin Mellen Press.

Chapter 5

Individual Scientist: Reasoning by the Numbers

5.1 INTRODUCTION

The teaching of statistical reasoning is, like that of reading and writing, part of forming an educated citizen.

Gigerenzer (2000)

Statistical inference emerged as the gold standard of objective scientific inference during the last century. As a career-long believer in statistical analysis, I acquired the habit of periodically reading outside the natural sciences to gain a better understanding of how other disciplines view and apply statistical techniques. This desire led me to publications by the German psychologist, Gerd Gigerenzer. In several of his publications (e.g., Gigerenzer, 2004, 2008), he described unsettling surveys by Haller and Krauss (2002) and Oakes (1986) that highlighted a fundamental misunderstanding by German psychologists including those who taught statistics, applied statistics in their research, and students who had passed one or more statistics courses. Haller and Krauss' survey presented the following results from a t-test of equal means: $t = 2.7$, $df = 18$, $P = 0.01$. Six misinterpretations of the associated P-value were then provided, and responders were asked to indicate whether each interpretation was correct or incorrect. The percentages endorsing one or more incorrect statements of the surveyed psychology students (100%), academic psychologists (97%), psychology scientists not teaching statistics (90%), and psychology methodology instructors (advanced graduate students) teaching statistics (80%) indicated pervasive confusion about what this fundamental metric signified. These results were so inconsistent with my comfort level with statistical inference that I suspected a bias in the survey itself. Acquiescence could bias the survey because responders were only provided wrong answers and would naturally try to accommodate the questioner by picking something under the assumption that a true answer was among the choices.

Given this suspicion and an unwillingness to automatically accept the disconcerting results as also relevant to environmental scientists, I administered a modified version of their survey to environmental science professionals and students at the following venues: southern China (57 students in a 2009

The Nature and Use of Ecotoxicological Evidence. DOI: https://doi.org/10.1016/B978-0-12-809642-0.00005-4

advanced course, Eutrophication and Environmental Risk Assessment, at Xiamen University), the United States (24 attendees of a 2009 Environmental Science Department seminar at the College of William & Mary, Virginia Institute of Marine Science; 27 attendees of a 2014 invited seminar at the University of Georgia's Savannah River Ecology Laboratory), central China (25 graduate students in a 2010 Practical Environmental Statistics Course at Huazhong Normal University), Spain (180 attendees of the 2010 SETAC Seville International Meeting), and southern India (61 attendees of the 2011 India Erudite Program lecture at Cochin University of Science and Technology). The occasions of this P-value survey were the same as those described in Chapter 2, Human Reasoning: Everyday Heuristics and Foibles, regarding poor estimation of pollutant presence with the only difference being that the survey was also done during a second seminar in the United States. Below are the question and answer choices. Unlike the Haller & Krauss (2002) survey, the correct answer was included as a choice and the instructions asked that only one best answer be selected.

Before reading further, you may wish to select the correct answer yourself. If you pick incorrectly, you can always reason that the question asks what you were first taught a "statistically significant" P-value meant, not your current understanding of its meaning.

> *When you took statistics, or when first exposed to significance testing, you were taught that a "statistically significant"" outcome with a p-value of 0.01 means . . .*
>
>   *(Check one answer)*
>   ☐ *1. The null hypothesis is disproven.*
>   ☐ *2. The p-value is the estimated probability of the null hypothesis being true.*
>   ☐ *3. You proved that an effect is present.*
>   ☐ *4. You can deduce the probability of the effect actually being present.*
>   ☐ *5. You know the probability that you would make a wrong decision if you rejected the null hypothesis.*
>   ☐ *6. You know the probability of getting these [or more extreme] data if the null hypothesis were true.*
>   ☐ *7. If you repeated the test many times, you would obtain a significant effect in 99 out of 100 trials.*

To my dismay, meta-analysis of the results (Fig. 5.1, Table 5.1) was consistent with the conclusions of Oakes (1986) and Haller and Krauss (2002). Only 9.6% (95% confidence interval (CI): 6.9%−13.2%) of those surveyed chose the correct answer (Answer 6). Randomly picking any answer from the seven choices would have resulted in 14.2% correct answers. This 14.2% is outside of the CI for the observed data, suggesting that the surveyed environmental science faculty and students were not simply poorly informed. They were misinformed during their training about the interpretation of P-values.

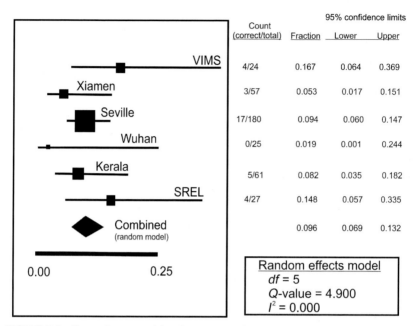

The following data accompanies the figure:

	Count (correct/total)	Fraction	95% confidence limits Lower	95% confidence limits Upper
VIMS	4/24	0.167	0.064	0.369
Xiamen	3/57	0.053	0.017	0.151
Seville	17/180	0.094	0.060	0.147
Wuhan	0/25	0.019	0.001	0.244
Kerala	5/61	0.082	0.035	0.182
SREL	4/27	0.148	0.057	0.335
Combined (random model)		0.096	0.069	0.132

Random effects model
$df = 5$
Q-value = 4.900
$I^2 = 0.000$

0.00   0.25

FIGURE 5.1   Forest plot summarizing the survey results (number selecting the correct answer and total number of answers) from six occasions. The square symbol indicates the fraction providing the correct answer with the symbol size reflecting the number of responses. The Count and Fraction columns reflect the number of correct response/total number of responses and the fraction of all responses that were correct, such as SREL had 4 correct responses of 27 responses or 0.148. A random model was applied although a fixed model might have sufficed because the df of 5 was very close to the Q-value (Borenstein, Hedges, Higgins, & Rothstein, 2009)). The I^2 value indicates that a trivial amount of the variation was associated with differences among occasions. The diamond symbol displays the combined results with its left and right points being at the lower and upper 95% confidence limits for the weighted estimate. VIMS, Virginia Institute of Marine Science in 2009; SREL, University of Georgia's Savannah River Ecology Laboratory in 2014; Xiamen, Xiamen University in 2009; Seville, SETAC Seville International Meeting in 2010; Wuhan, Huazhong Normal University in 2010; and Kerala, Cochin University of Science and Technology in 2011.

How could this be the case given the central role played by P-values in environmental sciences? Are there other statistical misconceptions that degrade our ability to make sound decisions and judgments based on statistical analyses? Perhaps misconceptions reflect a *status quo* bias or a misdirected informational cascade. This chapter explores these and related questions.

5.2   QUANTITATIVE METHODS

We live in a system of approximations.

Emerson (1844)

TABLE 5.1 Meta-Analysis Summary of Proportion of Responses to the Question "When you took statistics, or when first exposed to significance testing, you were taught that a 'statistically significant' outcome with a P-value of 0.01 means..." Answer 6 is the Correct Answer

Answer	Frequency	95% CI
1	0.078	0.042−0.141
2	0.234	0.184−0.292
3	0.107	0.052−0.208
4	0.096	0.070−0.132
5	0.205	0.114−0.341
6	0.096	0.069−0.132
7	0.209	0.116−0.349

A random effects model was used for all meta-analyses.

5.2.1   Explicitness of Quantitative Methods

As already described in Chapter 3, Human Reasoning: Within Scientific Traditions and Rules (Section 3.2.1), Popper (1959) favored hypotheses expressed in quantitative terms over those framed in qualitative terms. Quantitative hypotheses were more explicit, and consequently, more easily falsified during testing. Several 20th century pioneers in the emerging field of statistics were influenced by the advantages of falsification of quantitative hypotheses as they developed ways of implementing rigorous, notionally objective, tests with evidence.

As alluded to in Chapter 3, Human Reasoning: Within Scientific Traditions and Rules, the current integration of statistics into the scientific method includes the following general steps:

- Identify an unresolved problem or question.
- State a conceptual hypothesis that might contribute insight about that question.
- Based on the conceptual model, formulate a testable statistical null hypothesis that can potentially be falsified and used to make inferences about the conceptual hypothesis.
- Design an experiment to test the statistical hypothesis.
- Conduct that experiment and analyze the results with a null hypothesis significance test (NHST).
- Make evidence-based conclusions about the conceptual hypothesis from the NHST outcome.

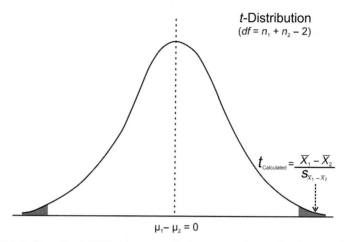

FIGURE 5.2   Example of NHST using a two-sided t-test assuming equal variances and sample sizes of n_1 and n_2. The H_{null} is that the means are equal ($\mu_1 - \mu_2 = 0$). Rejection regions are set at the tails of the t distribution PDF (gray shaded areas). In this case of a symmetrical PDF with a total area under the curve of 1, 0.025 of a total area is contained in each of the two rejection regions: there is a 0.05 probability of being in either rejection region by chance alone. A t-statistic is calculated from the sample means ($\overline{X}_1, \overline{X}_2$) and pooled standard deviation ($s_{\overline{X}_1 - \overline{X}_2}$), and compared to the rejection regions. The H_{null} of no difference in means is rejected if the calculated t-statistic is contained in either rejection region. The basis for rejection is a probability less than 0.05 of getting a calculated t-statistic of that (or greater) magnitude by chance alone. Notice that the rejection regions are the areas associated with 0.025 *or more extreme* calculated t-values. Put into terms of observations, the P-value is the probability of getting the observed data, or more extreme data, if the H_{null} were true.

Most implementations of NHST involve this general progression toward an evidence-based conclusion. An acceptable probability (α) of mistakenly rejecting a true null hypothesis (H_{null} or H_0) is established *a priori*. A distributional model is chosen based on the nature of the data, such as a t-statistic distribution if testing the difference between means of two unpaired samples with equal variances (Fig. 5.2). In the example shown in Fig. 5.2, a two-sided test might be chosen if a negative or positive difference in means is plausible. Traditionally, rejection regions were established at the two tails of the distribution (e.g., gray regions in Fig. 5.2) and the test was judged to be significant if the test statistic generated from the observed data (e.g., calculated t-statistic) were in either rejection region. The most common approach assumes a H_{null} of no difference and generates rejection regions accordingly. With the advent of computers and convenient software, establishing rejection regions has mostly been replaced by direct calculation of P-values from the assumed distribution probability density function (pdf). In the example here, the probability of getting the observed, or a more extreme, difference by chance alone (P-value) is derived from a t-statistic pdf with a defined degrees of freedom (*df*). (So Answer 6 is correct, and Answers 2 and 7 in the opening question are clearly incorrect). The H_{null} of no difference is rejected if *P*-value $<\alpha$. By convention, the α is most often fixed at 0.05 although some researchers report ranges of *P*-values to suggest the strength of the evidence against the H_{null}. For example, $P > 0.05$ (nonsignificant), $0.01 < P \leq 0.05$ (significant), or $P \leq 0.01$ (highly significant). The final step involves the "usual reject-H_{null}-confirm-the-theory" (Cohen, 1994, 1995) during which an alternative hypothesis ($H_{Alternative}$, such as a true difference in means of the two sampled populations) is accepted if the H_{null} is rejected. This approach is so conventional that Gigerenzer, a social scientist, considers it to be a social ritual.

Rituals seem to be indispensable for self-definition of social groups ... and there is nothing wrong with them. However, they should be the subject rather than the procedure of social sciences. Elements of rituals include (i) the repetition of the same action, (ii) a focus on special numbers or colors, (iii) fears about serious sanctions for rule violations, and (iv) wishful thinking and delusions that virtually eliminate critical thinking... the null [hypothesis significance testing] ritual has each of these four characteristics: incremental repetition of the same procedure; the magical 5 percent number; fear of sanctions by editors or advisors; and wishful thinking about the outcome, the p-value, which blocks researcher's intelligence.

Gigerenzer (2008)

It is so entrenched in many sciences that others have referred to the practice in similarly harsh terms: "mindless" (Bakan, 1966; Gigerenzer, 2004), "harmful" (Carver, 1978), "ritual" (Cohen, 1994; Germano, 1999; Guthery, 2008), "fallacy" (Goodman, 1999), "ill-founded" (Ioannidis, 2005), and "cargo cult science" (Guthery, 2008). The reasons for these disapproving descriptors are surveyed below.

5.2.2   Classic Statistical Testing and Inference

Although historically inexact[1], classic statistics is thought to have emerged in early 20th-century England largely due to the efforts and insights of Ronald Fisher, Francis Galton, William Gossett, Jerzy Neyman, Karl Pearson, and Egon Pearson. These statisticians developed and set in place many of the currently most popular statistical testing and regression methods. Agreements and conflicts emerging among these scholars were eventually integrated into an ostensibly cohesive collection of approaches; however,

---

1. Many central themes and approaches had already been established by the notables: Thomas Bayes, Daniel Bernuolli, Pierre-Simon LaPlace, Simon-Denis Poisson, Johann Carl Friedric Gauss, and Adrien-Marie Legendre. They include least squares estimation, curve fitting, Bayesian statistics, and the mathematical formulations for key distributions such as the Gaussian and Poisson distributions.

social dynamics resulted in inconsistencies that are important to understand. A prime example is the pervasive misunderstanding of P-values highlighted in the opening question of this chapter. A Kuhnian crisis (see Section 3.2.1) has emerged in statistics sciences as recent advances exposed major inconsistencies in conventional approaches and modern computers made possible the implementation of powerful alternatives. Books (e.g., Morrison & Henkel, 1970; Ziliak & McCloskey, 2007) and many journal articles (e.g., Cohen, 1994, 1995; Dixon, 2003; Gigerenzer, 2004; Wagenmakers, 2007) give testament to this crisis. The associated paradigm shift is only slowly diffusing outward from statistical science into sciences that rely on statistics when making inferences and decisions. Consequently, the basics of statistical inference from evidence are reexamined here with an emphasis on important points to keep in mind to avoid inferential errors. Modern alternatives are also highlighted.

5.2.2.1   Fisherian Significance Testing

The Fisherian approach has been the one almost universally used, most particularly by people who are not themselves statisticians and yet who use statistics, as in the social sciences. A cogent reason for the popularity of this approach is its ease of use, with standard formulas and computer software available.

Bauer (2012)

The Bayesian approach to statistical inference dominated statistical thinking prior to the early 20th century. Probabilities from the Bayesian vantage were measures of evidence-based credibility or plausibility of a hypothesis. Fisher (1958) referred to a Bayesian probability as a "specified state of logical uncertainty." Using evidence, investigators expressed their level of belief or certitude about hypotheses as probabilities.[2] The word belief inserted a subjective theme that frequentist statisticians would later consider inappropriate for scientific inquiry. The British statisticians mentioned above came to this conclusion and established a frequentist vantage as preferable for making sound inferences. The frequentists framed probabilities solely in the context of objective reality, such as the probability is 0.50 for an outcome of heads if a fair coin were flipped a very large number of times. This context for probability was independent of belief and depended solely on physical processes.

Ronald A. Fisher was the most prominent statistician to move the frequentist context to the center of statistical testing. Strongly influenced by the

---

2. Often the assertion of a prior probability was involved the Bayesian computations. Prior here means prior to the collection of the evidence or data. Depending on the circumstances, a prior probability might come from past experiments, the published literature, a hypothetical function, or a best guess. Coping with prior probabilities is a major challenge that frequentists often see as a major disadvantage of Bayesian methods.

inferential supremacy of falsification, he proposed an approach known as significance testing. Introducing the practical falsification or pseudo-falsification vantage, he asserted that a sufficiently improbable outcome from a statistical test can be treated as an impossible one. This practical assertion was made with full knowledge that it is not logically defensible because rare outcomes do occur. For example, despite the extremely low probability of winning a state lottery, someone does win. A P-value derived from a Fisherian significance test suggests whether the evidence is improbable enough that the associated H_{null} can be considered logically false. A false rejection of the H_{null} can occur so an acceptable rate or probability of making such an error is set *a priori* at some value (α).

The Fisherian significance test eventually became a central paradigm in statistical science and an impressive amount of normal science added details and refinements. As predictable from Thomas Kuhn's explanation of how scientific paradigms evolve and are eventually replaced, inconsistencies in significance testing slowly became apparent. Several key misconceptions associated with significance testing will be highlighted here.

First, the pseudo-falsification vantage asserts that a sufficiently improbable outcome can be treated as an impossible one and the H_{null} can be considered logically falsified. As pointed out by Cohen (1994), an error (inverse probability error) is committed in applying this logic. The probability of getting the evidence given that the H_{null} is true ($P(E|H_{null})$) is not the same as the probability of the H_{null} being true given the evidence ($P(H_{null}|E)$). Nor is there a consistent and straightforward correspondence between $P(E|H_{null})$ and $P(H_{null}|E)$. To derive one from the other requires more information than normally presented. Because the crux of the problem has already been described in the opening question of Chapter 2, Human Reasoning: Everyday Heuristics and Foibles, no further explanation will be given except to show that Bayes's Theorem explicitly makes this point,

$$P(H|E) = P(H)\frac{P(E|H)}{P(E)}.$$

In the significance testing context, the $P(E|H)$ is the probability of getting the evidence if the hypothesis were true. It is misinterpreted as being equivalent to $P(H|E)$, the probability of the hypothesis being true given the evidence. It can be stated generally that, if one misuses the P-value in this manner to suggest the probability of H_{null} given the evidence, it will consistently overstate the evidence against the H_{null}. That is, the H_{null} will be rejected more often than it should be at a specified error rate (Cohen, 1994; Newman, 2008).

Second, it is important to note that significance testing ostensibly involves logical falsification of one hypothesis, H_{null}. The probability of falsely (α) or correctly ($1 - \alpha$) rejecting the H_{null} is set before the experiment is conducted. Using evidence and test assumptions, one either rejects the H_{null} or fails to reject it. This means Answer 1 in the opening question cannot be correct because it is possible to falsely reject H_{null}. Despite the widespread practice of making inferences about a $H_{Alternative}$, Fisherian significance tests are focused on falsifying one hypothesis (H_{null}). No $H_{Alternative}$ exists so Answers 3 and 4 to the opening question which refer to a $H_{Alternative}$ are wrong. The "usual reject-H_{null}-confirm-the-theory" (Cohen, 1994, 1995) approach is invalid because no alternative hypothesis (theory) exists to be confirmed. Be that as it may, very few practitioners conduct significance tests without the hope of possibly confirming an alternative hypothesis, such as a significant difference in means exists.

Third, many significance tests are done with a H_{null} of no difference, that is, a H_{nil}. However, Fisher introduced the word, "null" to mean "the hypothesis to nullify," not "nil" as in "the hypothesis of no difference." Automatically applying a H_{nil} allows one to move ahead without putting much thought into deciding how large a difference is important. In most instances when a H_{nil} is invoked, ease of execution is chosen over thoughtful formulation of an insight-generating hypothesis. It is justified more by informational mimicry and ease of recall due to familiarity than sound reasoning. As you might recollect from Chapter 2, Human Reasoning: Everyday Heuristics and Foibles, mimicry is the best strategy under many conditions but can easily lead to maladaptive fads. It is unreliable if there is a high risk of using inappropriate or out-of-date information. I would argue that a Kuhnian crisis now exists and the risk of using obsolete information is high. Relative to familiarity, enough ecotoxicological ("local") knowledge now exists to set more informative H_{null}.

Fourth, Fisher states in numerous publications that he never intended the 0.05 level to become the default error rate. The error rate should be chosen based on the situation being studied and the stage of inquiry. Stigler (2008) speculates from historical records about Fisher's unintended contribution to the 0.05 convention. Fisher's watershed book, *Statistical Methods for Research Workers* (Fisher, 1925), required tables of statistics from the journal, *Biometrika*. Fisher had a long-standing intellectual feud with the journal editor, Karl Pearson, who refused to give permission for use of the tables. Fisher's only choice was to provide reduced tabulations of those statistics (e.g., for probabilities of 0.05, 0.02, or 0.01) most useful for the examples in his book. His landmark book also provided the first introduction of analysis of variance which would require tables for the many combinations of *F*-ratio numerator and denominator degrees of freedom. Too many *F*-statistic tables would have been required to accommodate all probabilities so Fisher chose again to generate only the tables most useful for his examples. Once the tact was taken of providing abridged tables for the inverse cumulative normal distribution and *F*-ratio, Fisher did the same for Gossett's *t*-distribution tables. Although expedient for producing his landmark book, the introduction of significance testing using reduced tables gave an unintentional false impression which still persists today.

Fifth, α is set *a priori* for practical falsification purposes so *post priori* categorizing of test results as nonsignificant, significant, or highly significant is nonsensical. Doing so is akin to saying "it is untrue," "it is true," or "it is really, really true." According to the logic of pseudo-falsification, it is treated as untrue if practically falsified or nothing can be concluded if it was not.

Sixth, the significance test ritual focuses on pseudo-falsification and gives short shift to the magnitude of the size of the effect. An ecotoxicologically trivial "significant" effect might easily be given more attention than an ecotoxicologically important, yet "insignificant," effect.

These and other reasons have been put forward for abandoning significance testing. The plausible counterargument is that these errors are made in the practice of significance testing and are not flaws in foundation concepts. For instance, Hurlbert and Lombardi (2009) recommend that the "paleoFisherian" approach just described and the about-to-be-described Neyman–Pearson paradigms be abandoned and a "neoFisherian" approach used in their place. (They also judge Bayesian methods as having overstated value and relevance.)

> *The essence of [the neoFisherian paradigm] is that a critical α (probability of type 1 error) is not specified, the terms "significant" and "nonsignificant" are abandoned, that high P values lead only to suspended judgments, and that the so-called "three-valued logic" of Cox, Kaiser, Tukey and Harris is adopted explicitly. Confidence intervals and bands, power analyses, and severity curves remain useful adjuncts in particular situations. Analyses conducted under this paradigm we term neoFisherian significance assessments (NFSA).*
>
> Hurlbert and Lombardi (2009)

Addressing the sixth misconception listed above, their NFSA further recognizes "the obvious, near universal need to present effect size information" (Hurlbert & Lombardi, 2009). Importantly, NFSA methods do not test hypotheses. Their three-valued logic refers to a broad approach espoused by Lehmann (1950), Kaiser (1960), Harris (1997), and others. As an example of this approach, the results of a two-tailed *t*-test might suggest that the researcher take one of three inferential positions: a difference exists in a positive direction, a difference exists in a negative direction, or no inference can be made either way from the evidence. Unlike some early advocates of three-valued logic, Hurlbert and Lombardi suggest that this approach be applied without requiring an *a priori* α. Although three-valued logic has not been widely adopted despite its long presence in the literature, Hurlbert and Lombardi judge that the current crisis in statistics makes it an appealing alternative. It remains to be seen whether the NFSA becomes generally accepted in statistical inquiry.

Responding to an increasingly sophisticated understanding of significance testing *P*-values and the associated questioning of the *status quo*, the American Statistical Association (Wasserstein & Lazar, 2016) recently recommended six points to be kept in mind to avoid *P*-value misconceptions and misapplications. They also attempt to move researchers away from the use of *P*-values in a pseudo-falsification framework.

- "*P*-values can indicate how incompatible the data are with a specified statistical model (such as a H_{null}) ... The smaller the *P*-value, the greater the statistical incompatibility of the data with the null hypothesis, if the underlying assumptions used to calculate the *P*-value hold."
- "*P*-values do not measure the probability that the studied hypothesis is true, or the probability that the data were produced by random chance alone ... [The *P*-value] is a statement about data in relation to a specified hypothetical explanation, and is not a statement about the explanation itself."
- "Scientific conclusions and business or policy decisions should not be based only on whether a *P*-value passes a specific threshold. Practices that reduce data analysis or scientific inference to mechanical 'bright-line' rules (such as '$P < 0.05$') for justifying scientific claims or conclusions can lead to erroneous beliefs and poor decision making."
- "Proper inference requires full reporting and transparency." *P*-values and related details should not be reported selectively as in the cases of reporting only "significant" *P*-values or not specifying how many statistical tests were conducted.
- "A *P*-value, or statistical significance, does not measure the size of an effect or the importance of a result. Statistical significance is not equivalent to scientific, human, or economic significance. Smaller *P*-values do not necessarily imply the presence of larger or more important effects, and larger *P*-values do not imply a lack of importance or even lack of effect."
- "By itself, a *P*-value does not provide a good measure of evidence regarding a model or hypothesis."

5.2.2.2 Nyman−Pearson Hypothesis Testing

Any discussion of an alternative hypothesis during Fisherian significance testing is misguided because none exists. Nonetheless, discussions of $H_{Alternative}$ are pervasive in interpretations of significance tests because the early coming together of significance and hypothesis testing approaches uncritically blended the alternative hypotheses used in Nyman−Pearson hypothesis testing into Fisherian significance testing.

Jersy Neyman and Egon Pearson, the son of Karl Pearson, believed that Fisher's significance testing which aims at logically falsifying hypotheses was not useful to researchers trying to decide which of two hypotheses was most probable given the evidence. They formulated the hypothesis testing method that involved two error rates, Type I and II. The Type I error rate, usually designated α, is the probability of falsely rejecting the primary (commonly termed null) hypothesis. The Type II error rate (β) was the probability of falsely rejecting the secondary (commonly termed alternative) hypothesis. Importantly in

Neyman–Pearson hypothesis testing, the seriousness of making a Type I or II error dictates what values of α and β should be used. Often the ratio, α/β, is applied as a rough guide to the relative seriousness of these two errors.

> *When the errors can be distinguished by their gravity, the more serious of them is normally called Type I error ... suppose two alternative theories concerning a food additive were entertained, one that the substance is safe, the other that it is highly toxic ... it would be less of a danger to assume that a safe additive was toxic than a toxic one was safe.*
>
> Howson and Urbach (1989)

The intent of hypothesis testing, in contrast to significance testing, is to help the researcher decide which hypothesis would be the best bet in the long run. Unlike Fisher's significance testing, it does not attempt to discern which hypothesis was logically false. It simply provides a way for deciding from evidence which hypothesis is most likely given the evidence.

> *Unlike Fisher's approach, the Neyman-Pearson approach aims only to guide future behavior about the proposed hypotheses (i.e., to act as if one or another hypothesis were true), not to infer from the experiment that a null hypothesis was falsified. According to the Neyman-Pearson line of reasoning, you are more likely to be correct in the long run if you behave toward a hypothesis in a manner suggested by the test results.*
>
> Newman (2008)

Setting α and β requires specifying a meaningful effect size (ES) to be detected during testing if present. Often involving a pilot study or thoughtful literature analysis of similar studies, ES estimation is seen as a burden by some researchers; however, deciding what constitutes a meaningful ES must eventually be done to judge the importance of research results. Once α, β, and ES are established, the necessary number of replicates or samples is estimated by conventional sample size or power analyses.

In contrast to the two possible outcomes of significance testing (reject or do not reject the H_{null}), a completed hypothesis test has four possible outcome states:

1. Reject a true H_{null}: Type I error, α.
2. Accept a true H_{null}: $1 - \alpha$.
3. Reject a true $H_{Alternative}$: Type II error, β.
4. Accept a true $H_{Alternative}$: Power, $1 - \beta$.

Again, notice that Answers 3 and 4 to the opening question are wrong because the question was framed as a Fisherian significance test which has no $H_{Alternative}$ of an effect being present. There would have been an $H_{Alternative}$ if the Neyman–Pearson hypothesis testing context had been presented with specified α, β, and ES.

5.2.2.3   Useful Embellishments and Extensions

Several satellite concepts surround significance and hypothesis testing. The NFSA has already been mentioned as one associated with significance testing. Three more deserving mention are pertinent to the Neyman–Pearson hypothesis testing paradigm.

Mudge, Baker, Edge, and Houlahan (2012) point out that the arbitrary setting of α has come under increasing scrutiny and propose that an optimal α be picked instead that minimizes Type I and II errors for any specified ES. They provide a straightforward illustration of how to calculate such an optimal α. This approach is laudable but the balancing of error rates based on the relative seriousness of making each type of error is not incorporated.

Adaptive inference (Holling & Allen, 2002) is another refinement presented as more relevant than strong inference (Platt, 1964) in fields such as ecology. In contrast to physics or many laboratory-centered sciences, ecology often explores phenomena with potentially multiple causes and confounding factors. As a consequence, it would be a rare instance in which a sequence of discriminating tests methodically isolated a single cause for an ecological effect. An adaptive inference approach is advocated instead that emphasizes shifts in Type I or II error rates as the ecologist moves from the initial stages of inquiry with high uncertainty toward the more informed, later stages of inquiry. Holling and Allen's approach has been modified in Fig. 5.3 to better reflect the conditions and intentions of ecotoxicologists and ecological risk assessors. The chance of prematurely rejecting sound hypotheses during initial testing is decreased by adjusting the Type II error rate downward to favor high power. Initially, the worst error to make would be to prematurely reject a sound $H_{Alternative}$. As uncertainty is reduced by continued accumulation of evidence from increasingly more definitive experiments, the emphasis would shift toward minimizing the probability of a Type I error, that is, mistakenly accepting an unsound hypothesis. The error rates shift from ensuring that good hypotheses are not prematurely discarded to ensuring that poor hypotheses are not mistakenly accepted later in the research program. Despite the obvious advantage of this approach, it would be difficult, given the present day pressure to publish, to avoid publication of the initial investigations before the entire investigative process had run its course.

The final embellishment is to estimate of the probability of the $H_{Alternative}$ being true given a significant hypothesis test outcome (Doğan & Doğan, 2017; Ioannidis, 2005). This probability is called the positive predictive value (PPV),

$$PPV = \frac{(1 - \beta)R}{R - \beta R + \alpha}.$$

The R needed to calculate the PPV is the quotient, "true relationships"/ "no relationships" estimated, before testing. As will be described more

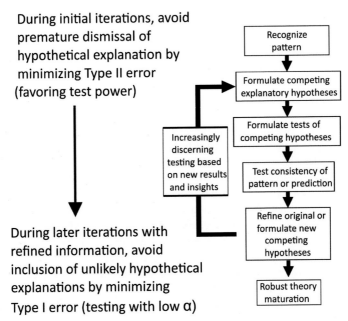

During initial iterations, avoid premature dismissal of hypothetical explanation by minimizing Type II error (favoring test power)

During later iterations with refined information, avoid inclusion of unlikely hypothetical explanations by minimizing Type I error (testing with low α)

Recognize pattern

Formulate competing explanatory hypotheses

Increasingly discerning testing based on new results and insights

Formulate tests of competing hypotheses

Test consistency of pattern or prediction

Refine original or formulate new competing hypotheses

Robust theory maturation

FIGURE 5.3 The Holling and Allen (2002) adaptive inference context modified to be directly pertinent to ecotoxicology and environmental risk assessment *Modified from figure 1 in Holling, C.S., & Allen, C.R. (2002). Adaptive inference for distinguishing credible from incredible patterns in nature. Ecosystems, 5, 319–328.*

precisely in Section 5.2.5, R is the Bayesian prior odds of a true relationship to no relationship. It follows that the prior probability of a true relationship is $R/(R + 1)$. Past experiments, expert elicitation, or a literature search might be used to estimate R. If no information is available to estimate R, the principle of indifference could be applied, that being, assume there is an equal chance of one or the other condition ($R = 1$).

Also useful is the negative predictive value (NPV), that is, the probability of the null hypothesis being true given a nonsignificant hypothesis test result.

$$NPV = \frac{1 - \alpha}{1 + \beta R - \alpha}.$$

Now, probabilities of most interest to researchers can be estimated from test results (Fig. 5.4). These straightforward relationships make it clear that Answer 5 to the opening question was wrong because the provided details were insufficient for estimating PPV. Nonetheless, 1 out of 5 surveyed environmental scientists (proportion of 0.205 in Table 5.1) misinterpreted the *P*-value as being the false-positive result probability, FPRP = 1 − PPV. Application of PPV and NPV is well established in many fields (Rizak &

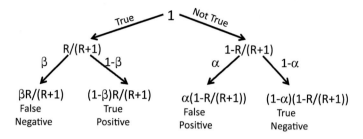

$$PPV = \frac{True\ Positive}{True\ Positive + False\ Positive} = \frac{(1-\beta)R/(R+1)}{(1-\beta)R/(R+1)+\alpha(1-R/(R+1))}$$

$$NPV = \frac{True\ Negative}{True\ Negative + False\ Negative} = \frac{(1-\alpha)(1-R/(R+1))}{(1-\alpha)(1-R/(R+1))+\beta R/(R+1)}$$

FIGURE 5.4 Graphic illustration of PPV and NPV for testing that a hypothetical relationship is either true or false. R, the number of true relationships divided by the number of false relationships, is the estimated prior of a true relationship. It follows that P(True) is $R/(R+1)$ and P(False) is $1 - R/(R+1)$. Two additional probabilities can be estimated with this approach. The false-positive result probability (FPRP) is $1 - $ PPV and the false-negative result probability is $1 - $ NPV (Newman, 2013).

Hrudey, 2006; Wacholder, Chanock, Garcia-Closas, El Ghormli, & Rothman, 2004) and has been discussed relative to exploring statistical test results in ecotoxicology (Bundschuh et al., 2013; Newman, 2008, 2013).

> *In view of the prevalent misuses of and misconceptions concerning p-values, some statisticians prefer to supplement or even replace p-values with other approaches. These include methods that emphasize estimation over testing, such as confidence, credibility, or prediction intervals; Bayesian methods; alternative measures of evidence, such as likelihood ratios or Bayes Factors; and other approaches such as decision-theoretic modeling and false discovery rates.*

> Wasserstein and Lazar (2016)

Explorations of still other alternative approaches have been prompted by inconsistencies in the practice of significance and hypothesis testing, especially misunderstandings about *P*-values, hypotheses, and error rates. Estimation with intervals and model comparison with likelihood ratios or information criteria are emerging as plausible augmentations or replacements of conventional statistical tests. The previously disparaged Bayesian approach is also gaining considerable support. The rapid increase in computational resources has accelerated the exploitation of Bayesian and information-theoretic approaches.

5.2.3 Confidence Intervals

Recent statistical traditions emphasized P-values and test significance to such an extent that other important features were ignored, resulting in the loss of valuable insights. As a very relevant example, Anderson, Burnham, and Thompson (2000) found that 47% of P-values reported in *The Journal of Wildlife Management* from 1994 to 1998 were naked, that is, "only the p-value is presented with a statement about its significance, without estimated effect size or even the sign of the difference being provided" (Anderson, Link, Johnson, & Burnham, 2001). Estimation including CIs is an appealing alternative to formal statistical testing that makes consideration of ES more convenient (Gardner & Altman, 1986).

The virtue of CIs is that they simultaneously provide a convenient way to understand ES and its associated precision, while moderating the overemphasis on statistical significance (Anderson et al., 2001). In the case of the *British Medical Journal*, their advantage over statistical testing was so obvious that papers applying statistical testing were automatically rejected at one point and authors were encouraged to use confidence limits instead (Altman, Machin, Bryant, & Gardner, 2000; Fidler, Thomason, Cumming, Finch, & Leeman, 2004).

Application of CIs is readily acceptable to most for several reasons. Confidence intervals are taught during conventional statistical courses and familiarity makes most people comfortable with their use. Equally important, confidence limits allow simultaneous presentation of ES, uncertainty in estimates, and, if desired, the statistical significance of the results (Di Stefano, 2004; Nakagawa & Cuthill, 2007; Poole, 2001). Emphasis on size effect is a great advantage of CIs given its habitual neglect during statistical testing. For the researcher who wishes to link CIs to significance tests, P-values can be calculated or general overlap rules used such as the following for two independent means (Cumming, 2012; Cumming & Finch, 2005).

1. If the 95% CIs on two independent means **just touch** end-to-end, overlap is zero and the P-value for testing the null hypothesis of no difference is approximately 0.01.
2. If there's a **gap** between the two *CIs*, meaning no overlap, then $P < 0.01$.
3. **Moderate overlap** (Moderate overlap of two confidence intervals of similar lengths is about one half of one of the CI arm lengths.) . . . of the two CIs implies that P is approximately 0.05. Less overlap means $P < 0.05$.

Cummings (2012)

The widely known US EPA fathead minnow larval survival data set (Weber et al., 1989) can be used to demonstrate the value of CIs (Fig. 5.5). With a t-test, the conclusion from these data might have been that a significant difference exists between the 0 and 512 µg/L sodium pentachlorophenol treatments. With the CI approach, a significant

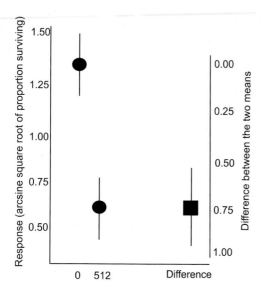

FIGURE 5.5 Comparison of mean responses for fathead minnow larvae exposed to 0 and 512 µg/L sodium pentachlorophenol. Round symbols with their error bars indicate treatment means and 95% CIs. The difference between the unpaired means and its associated 95% CI are indicated with the square symbol and its error bars. The right axis is that for the difference between treatment means; the left axis is the response observed (arcsine square root of the proportion of exposed larvae dying).

difference is immediately obvious by applying the above general overlap rule. Also zero is not within the CI for the difference in means as shown on the right side of the figure. The precision (length) of the CIs suggests adequate sample sizes for comparing treatment means. It is also immediately obvious that larval survival was substantially reduced by the presence of sodium pentachlorophenol.

Still more involved analyses can be done with CIs as detailed in publications such as Altman et al. (2000) and Cumming (2012). For example, Fig. 5.6 depicts the analysis of the entire pentachlorophenol data set that includes an exposure concentration series. Conventional analysis would involve Dunnett's test because multiple comparisons are being done and adjustment of pairwise Type I error rates is thought to be necessary to maintain a specific experiment-wise error rate. As shown in the figure, these same error rate adjustments can be done to produce Dunnett's simultaneous CIs (Delignette-Muller, Forfait, Billoir, & Charles, 2011). Survival in this example was substantially affected in the highest treatment with survival being less than half of the 0 µg/L treatment. The 95% CI for the associated ratio did not overlap with 1. If, based on knowledge of the species, a decrease of 10% or more in survival was judged to be demographically unacceptable, the magnitude of the decrease for the 256 µg/L exposure and its relatively wide CI would also suggest that more information is needed before concluding that survival at that second highest concentration is not a concern.

Two factors impede the more effective use of estimation with CIs. One is a general misconception and the other is a prevailing attitude in the scientific community.

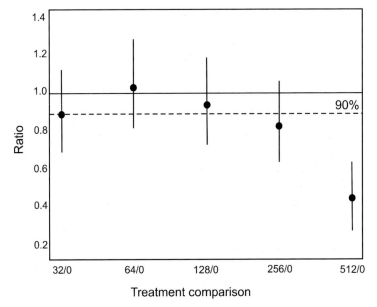

FIGURE 5.6   Dunnett's simultaneous 95% CIs for fathead minnow larvae survival after exposure to sodium pentachlorophenol. Confidence intervals for the ratio of survival effects are used.
The pair of treatments being used to construct CIs is shown on the x-axis. For example, 32/0
means the ratio of survival in the 32 μg/L treatment to that in the 0 μg/L treatment. Absolutely no
effect of treatment would result in a ratio of 1 (black horizontal line). Here a dashed line is added
to indicate a (speculative) meaningful difference that, if exceeded, would result in harm to fathead
minnow populations. Using the CIs, it is clear that the 512 μg/L treatment results in a statistically
and biologically significant effect. (Computations were done with the R mratios package.)

A general misconception exists about what exactly CIs are. Assume that
a sample is taken from a population and a parameter—often the mean—is to
be calculated from that sample. If the sampling and parameter CI calculation
process were repeated many times, 95 of 100 of the resulting intervals would
contain the true population mean (Simm & Reid, 1999). Although often
done, it is incorrect to say of any particular calculated CI that there is a 95%
chance that the true mean is within its confidence limits (Cumming, 2012).

The prevailing attitude that CIs are inferior to formal statistical tests can
be illustrated with an instance of a review of one of my own papers.
Confidence intervals were intentionally used instead of conventional statistical tests in the paper (Newman, Xu, Cotton, & Tom, 2011) as a personal
experiment to see how this would impact the review process. I predicted that
discomfort would be produced by any paper that did not apply conventional
testing. Consistent with this prediction, two of the three reviewers pointed
out that the lack of conventional statistical testing was a serious flaw.
Reviewer 1 stated, "I would have expected that the author [would] have

made better use of statistics in backing the correlations—I am not sure that the inferences in the paper would stand up to a [rigorous] statistical analysis." The reviewer also misinterpreted the 95% CIs as being standard errors of the mean, stating that the standard errors were suspiciously wide. Reviewer 2 opined, "... on the downside, the use of simple comparisons of 95% CIs to compare, contrast and correlate data rather than more rigorous statistics significantly reduces the apparent quality of the work." The editor rejected the paper. With the submission experiment over, I emailed the editor to acknowledge rejection, thank him for his work, and mention growing concerns about statistical significance testing. I attached to the email several papers by others who advocated replacement of statistical testing by CIs whenever possible. Going beyond expected diligence, the editor carefully read the papers and reversed his decision provided that a brief explanation be inserted about why CIs were used instead of conventional tests. The implication from this exercise is that there is resistance to publications not using "rigorous" conventional statistical testing but the manuscript vetting process appears open to arguments for applying CIs instead.

5.2.4 Information-Theoretic Methods

The name [information-theoretic] comes from the fact that the foundation [for this approach] originates in "information theory"; a set of fundamental discoveries made largely during the World War II with many important extensions since that time. One exciting discovery is the ability to actually quantify information and this has led to countless breakthroughs that affect many things ...

Anderson (2008)

Information theory emerged during World War II in efforts to deal with the information content of signals and encrypted materials. Kullback and Leibler (1951) generated the seminal paper on this approach so any related concepts are often referred to as Kullback-Leibler information theory. Kullback and Leibler developed formulae for estimating the loss of information of a "reality" or true situation when applying some approximating model (Anderson et al., 2000). Like the CI approach, the information-theoretic approach emphasizes inference from estimation, not significance testing. What is distinct about this approach is its use of measures of information content in models (hypotheses) to select the most useful of candidate models. In this approach, plausible models are hypothesized and then assessed with metrics of information content. The favored model has the least loss of information relative to that contained in the "reality" being studied (Anderson, 2008). An example involving two hypothesized models might be that a set of five observations is best described by either a normal or lognormal distribution (Fig. 5.7). As reflected in the associated likelihoods ($L(H_N)$ and $L(H_{LN})$), the better fitting model will retain more of the information

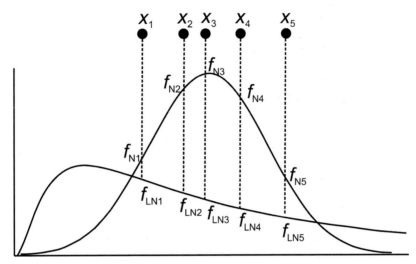

FIGURE 5.7 Example of model selection framed in terms of a likelihood ratio. Five observations (X_1-X_5) are available and two model distributions proposed for the population from which the samples were taken. The probability densities (f_{N1-5}, f_{LN1-5}) from each model distribution for each point can be multiplied together to produce a likelihood of obtaining the observations from the two competing distributions. Because both models have the same number of estimated parameters, no further calculations are needed to determine which model reflects the least amount of information loss. The normal distribution model seems best here.

associated with the "real" population being sampled. (Recollect that the likelihood is the probability of the set of observations as a function of the model. For instance, the $L(H_N)$ is the probability of the five observations under the assumption of a normal model with an estimated mean and variance. More explicitly, the likelihood is "*any constant multiple* of the probability of the observed data..." (Aitkin, 2010)). The normal model might be favored if the information loss associated with it was less than that for the lognormal model. In the simple example shown in Fig. 5.7 in which both candidate distributions have the same number of estimated parameters, a simple likelihood ratio can be used to gauge which is best,

$$\frac{L(H_N)}{L(H_{LN})} = \frac{\prod_{i=1}^{5} f(x_i|H_N)}{\prod_{i=1}^{5} f(x_i|H_{LN})}.$$

More complex models might serve as hypotheses and measures of information content for each can then be used to select the most appropriate of the models. Likelihoods alone are insufficient in cases for which candidate models have different numbers of parameters. All else being equal, the likelihood will improve as more and more parameters are added to any model.

Some metric incorporating fit and number of parameters must be used instead. "Models with too few parameters (variables) have bias, whereas models with too many parameters (variables) may have poor precision or tend to identify effects that are, in fact, spurious ... These considerations call for a balance between under- and over-fitted models..." (Burnham & Anderson, 2001). Likelihood for a model can be adjusted with a metric such as the Akaike Information Criterion (AIC). The AIC adjustment of a model's likelihood (L(Model|Observations)) for the number of parameters in a model (K) is very straightforward, AIC $= -2\ln(L(\text{Model|Observations})) + 2K$. The L(Model|Observations) in the AIC is the maximized likelihood of the model given the observations. In many cases, there is a small sample size bias in AIC so an adjusted AIC_c is calculated instead (Anderson et al., 2001; Johnson & Omland, 2004).

The AIC or AIC_c are compared for candidate models and the model with the smallest information criterion is the one with the least amount of information loss. Another way of envisioning this minimum AIC estimation (MAICE) process is in terms of the Principle of Parsimony. MAICE allows one to select the model with the most information content per estimated parameter.

5.2.5 Bayesian Inference

Before the 1920s, Bayes's Theorem was widely regarded as the proper basis for inductive reasoning. However, Fisher argued forcefully that it could not deliver completely objective appraisals of hypotheses ... and he led a highly influential and effective campaign against the Bayesian approach.

Howson and Urbach (1989)

For many years, statistics textbooks have followed this 'canonical' procedure: (1) the reader is warned not to use the discredited methods of Bayes and Laplace, (2) an orthodox method is extolled as superior and applied to a few simple problems, (3) the corresponding Bayesian solutions are not worked out or described in any way. The net result is that no evidence whatsoever is offered to substantiate the claim of superiority of the orthodox method.

Jaynes (1976)

Up until very recently, the frequentist context overshadowed Bayesian interpretation of probabilities as expressions of the degrees of credibility in the light of evidence (Howson & Urbach, 1989). Dominance of the frequentist vantage was achieved through the argument that only objective information and no subjectivity should be involved in scientific inferences. However, a counterargument is now put forward with increasing frequency that the selection of error rates, ES, hypotheses, acceptable assumptions, and applied tests introduce subjectivity into frequentist methods. Also, difficult computations that made many Bayesian applications arduous have been resolved with computer-intensive procedures. As an important example, some common Bayesian methods had required complex integrations that can now be approximated with Markov Chain Monte Carlo (MCMC) sampling (Lesaffre & Lawson, 2012).

The difficulties of making frequentist inferences about $P(H|E)$ from $P(E|H)$, the direct Bayesian inferences possible about $P(H|E)$, and availability of computer-intensive techniques have led to a resurgence of Bayesian statistics. Many of the difficulties mentioned above can be resolved with Bayesian methods. However, it is important at the onset of this treatment to point out that many of these resolutions come at a price. Many Bayesian methods require prior probabilities (also known as "priors") that can be challenging to produce. This legitimate shortcoming of Bayesian analysis is commonly invoked by individuals resistant to abandoning frequentist methods. Nonetheless, numerous introductory Bayesian statistics textbooks (e.g., Aitkin (2010), Bolstad (2007), Kruschke (2011), Lesaffre and Lawson (2012), and Woodworth (2004)) are now available and supported by a rapidly increasing number of excellent computer packages, especially R packages. Most conventional statistical methods now have convenient Bayesian equivalents.

The diachronic Bayes's theorem already presented several times elsewhere in this book can be used as the starting point for describing Bayesian inference from evidence.

$$P(H|E) = P(H) \frac{P(E|H)}{P(E)}$$

Two candidate hypotheses—say hypothetical distributions, H_N and H_{LN}—might be proposed as underlying distributions for a population from which observations (E) were taken. The quotient of the two probabilities, $P(H_N|E)$ and $P(H_{NL}|E)$, can be calculated with $P(H_N)$, $P(H_{NL})$, $P(E|H_N)$, and $P(E|H_{NL})$,

$$\frac{P(H_N|E)}{P(H_{NL}|E)} = \frac{P(H_N) \frac{P(E|H_N)}{P(E)}}{P(H_{NL}) \frac{P(E|H_{NL})}{P(E)}} = \frac{P(H_N)}{P(H_{NL})} \frac{P(E|H_N)}{P(E|H_{NL})}.$$

Notice that $P(E)$ conveniently cancels out and does not need to be estimated. The $P(H_N)/P(H_{NL})$ is the ratio of the two probabilities before the evidence, E, is collected. If no information was available to estimate this prior ratio, the principle of indifference can be applied. Perhaps, the ratio of prior probabilities, $P(H_N)/P(H_{NL})$, would be set to 1.[3] (If the focus is on

---

3. This simple example should not mislead the reader into thinking that prior ratio selection is easy. Selection of an appropriate prior can be quite involved and various approaches have emerged for estimation, even for those framed around the principle of indifference. See as an example the discussion in Rouder, Speckman, Sun, and Morey (2009) about selecting priors for Bayesian t-tests. Selection of priors can influence estimated Bayes factors (e.g., see pages 260–262 and 303 in Kruschke (2011) or page 30 in Lesaffre and Lawson (2012)) and credible intervals (e.g., see page 152 in Woodworth (2004) described below.

distributions, a uniform prior distribution might be selected initially in which all values have equal probabilities.) The $P(E|H_N)/P(E|H_{LN})$ is the ratio of the probabilities of the evidence under the assumptions of the hypotheses. It is also called the likelihood ratio. Notice that, if $P(H_N)/P(H_{NL})$ was 1, the relationship is the same as the one given above for the likelihood ratio.

$$\frac{P(H_N|E)}{P(H_{NL}|E)} = 1 \cdot \frac{P(E|H_N)}{P(E|H_{NL})} = \frac{L(H_N)}{L(H_{LN})}.$$

Hypotheses can be compared to one another using the likelihood ratio. Notice that, unlike the previously discussed information-theoretic application of likelihood ratios, Bayesian methods also provide the opportunity to integrate any available prior knowledge into the computations. The $P(E|H_1)/P(E|H_2)$ ratios are called Bayes factors (BF) in this context and their magnitudes are used to express the relative evidence-based credibility of competing hypotheses. In the above example, a BF of say 10 would mean that the ratio of the evidence-based probabilities of H_N to H_{NL} is 10 times larger when computed using the posterior rather than the prior (Anderson et al., 2001).

For convenience, BFs are often converted to minimum Bayes factors (mBFs) by placing the hypothesis with the least support in the numerator. That assures that the mBF is 1 or less. Using BFs to judge the relative credibility of hypotheses is based on a general scaling suggested by Jeffreys (1983). Table 5.2 summarizes one permutation of Jeffrey's scale (Goodman, 2001) in which the null hypothesis is placed in the numerator, mBF = $P(E|H_{null})/P(E|H_{alternative})$.

The sodium pentachlorophenol toxicity data for fathead minnows used previously to illustrate CI estimation can be used again to show how convenient mBFs are for making inferences (Fig. 5.8). The P-values for mortality at each concentration (32, 64, 128, 256, and 512 vs 0 µg/L NaPCP) from conventional t-tests (with and without Bonferroni adjustments) are depicted on the y-axis. The corresponding mBFs are depicted on the x-axis. The conventional t-test indicates a significant deviation in mean mortality in the 512 µg/L treatment relative to the H_{null} of no difference from the 0 µg/L NaPCP treatment. According to conventional misinterpretation of the results, the mean mortality in the 512 µg/L treatment (more than 50% mortality) was significantly higher than that in the 0 µg/L treatment but mortality at the other concentrations were not significantly different from that of the 0 µg/L treatment. This was the conclusion with (red region) or without (yellow region) Bonferroni adjustment of pairwise error rates. From the mBF context (x-axis), the evidence against the H_{null} was very strong for the 512 µg/L treatment but only weak for all other treatments. Importantly, inferences from the mBF are direct ones about the competing hypotheses, whereas those from the t-test are indirect ones prone to error.

A more involved application of inference based on BF can be illustrated with a recent laboratory study of mercury-contaminated sediment and two

TABLE 5.2 Evidence-based Credibility of the H_{null} based on mBF

mBF	Evidence Strength Against the H_{null}
0.26	Weak
0.15	Moderate
0.10	Moderate
0.04	Moderate to strong
0.005	Strong to very strong

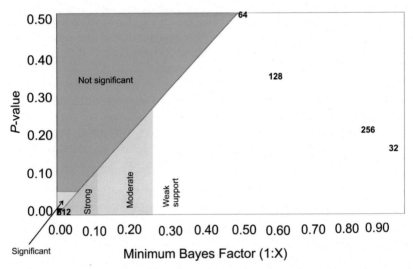

FIGURE 5.8 Comparison of P-values from conventional t-tests (with and without Bonferroni corrections) and mBFs for the sodium pentachlorophenol fathead minnow larvae toxicity data set (Weber et al., 1989). The numbers indicate the P-value (y-axis) and mBF (x-axis) for the comparison of mortality at that concentration treatment to the $0 \, \mu g/L$ treatment. Regions of P-values that would not be significant, be significant with an unadjusted t-test, and be significant with a Bonferroni adjusted test are indicated as green, yellow and red regions, respectively. The mBF regions of weak support, moderate, and strong support (against the H_{null} being true) are indicated in white, blue and lavender, respectively.

sorbent amendments on detrital processing by the amphipod, *Hyalella azteca*. The experimental design was similar to that of Bundschuh, Zubrod, Seitz, Newman, and Schulz (2011) and involved exposures to mercury-contaminated (Dooms Crossing, 8.1 μg Hg/g dry weight, 95% CI: 7.0−9.2, $n = 14$) and uncontaminated (North Oak Lane, 0.038 μg Hg/g dry weight, 95% CI: 0.036−0.039, $n = 10$) sediments. Sediments were left untreated or treated with a biochar or the commercial sorbent, Sedimite. Tests were done with sediments 0, 1.5, 3, and 6 months after sorbent addition to sediments to determine how long any sorbent might influence detrital processing. Individual amphipods fed for 10 days on leaf disks placed on the sediments and leaf weight loss quantified as a measure of detrital processing. Conventional sample size estimation from a pilot study indicated that 30 replicates/treatment was adequate to detect a 35% ES with α and β of 0.05. Interpretation of resulting mBF was based on Jeffrey's general classification of evidence strength (Fig. 5.9).

The mBF provided no evidence that amending contaminated or uncontaminated sediments with biochar influenced detrital processing (left most column, "Biochar vs Sediment"). This was true for all durations from 0 to 180 days post-sediment amendment. However, mBF provided strong to very strong evidence in six of the eight treatments of decreased detrital processing (relative to unamended sediment) with the addition of Sedimite (middle

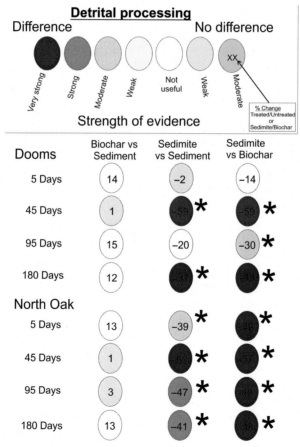

FIGURE 5.9 Analysis of amphipod detrital processing assay using mBF. For comparative purposes only, results from conventional t-tests are also shown with an asterisk (*) indicating a significant difference ($\alpha = \beta = 0.05$ at ES = 35%).

column, "Sedimite vs Sediment"). Decreases ranged from 2% to 53%. In seven of eight trials comparing detrital processing in the presence of Sedimite or biochar, mBF indicated strong to very strong evidence that Sedimite amendment resulted in much lower detrital processing rates than did biochar amendment (right most column, "Sedimite vs Biochar"). The decrease ranged from 14% to 59% with no evidence of change through time.

The roundabout inferences from CIs described above can be made straightforward by constructing Bayesian intervals for parameter estimates instead. Recollect that the interpretation of frequentist CIs was that 95% of such CIs in the long run would contain the true parameter value. For a Bayesian credible (or credibility) interval, the interpretation is that there is an evidence-based 95% probability that the true parameter value lies *within this specific interval*. Put another way, the credible interval includes a set of parameter estimates with a specified (usually high) credibility expressed as a probability. Credible intervals define the level of belief warranted by the specific in-hand sample of observations. Although often interpreted in this way, this is not what CIs do. That having been said, credible intervals calculations for straightforward situations are often (but certainly not always) similar to those of CIs.

Credible intervals are in many cases numerically very similar to what conventional, frequentist statisticians call confidence intervals; however, they are conceptually distinct, and only credible intervals describe the knowledge (beliefs) supported by the data that were actually observed.

Woodworth (2004)

Credible intervals can have equal tails (Fig. 5.10, top panel) with a 95% (posterior) probability interval being between 2.5% and 97.5%, or they can have different right and left tail percentages. In the latter case, they are referred to as highest (posterior) density intervals (HDI). The HDI will be the narrowest interval containing a 95% probability mass for a distribution. For some situations for which the assumption of equal tails is unreasonable, it might be necessary to generate HDIs with methods capable of estimating areas under complex distributions (Fig. 5.10) so that intervals containing 95% of the probability mass can be generated (Hyndman, 1996). Whether the interval is expressed as a credible or highest density interval, the context during inference is that parameter values outside of the interval are less credible than those inside the interval (Kruschke, 2011).

Bayesian equivalents of other statistical methods are also becoming commonplace. The information-theoretic AIC_c for model selection has its Bayesian equivalent in the Schwarz or Bayesian information criterion (Johnson & Omland, 2004). Sutton and Abrams (2001) describe Bayesian meta-analysis techniques like the frequentist meta-analyses used for the survey data in Figs. 2.1 and 5.1. The formal analysis of evidence such as that depicted in Fig. 3.4 can be facilitated by Bayesian belief networks (e.g.,

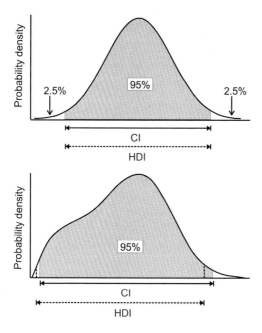

FIGURE 5.10 Credible intervals (CI) and highest density intervals (HDI) for two hypothetical probability density distributions of a parameter. A symmetrical distribution (top panel) will have CI and HDI that contain a set of credible parameter values in the central 95% of area of the curve. Intervals for more complex distributions (bottom panel) can be approximated with methods such as MCMC. The CI and HDI will not be the same in this situation. The HDI will be the narrowest interval containing 95% of the area and often will not have equal tails like the CI.

Carriger, Barron, & Newman, 2016; Neapolitan, 2004; Pearl, 2000; Scutari & Denis, 2015).

5.3 CONCLUSIONS

Widespread misapplication of NHST in ecotoxicology and environmental risk assessment can be easily illustrated (Fig. 5.11). Extending an earlier literature analysis (Newman, 2008), results from a sequence of three environmental science and chemistry journal surveys (ETC 1: 1996−2006; ETC 2: 2007−11; ETC 3: 2012−16) were compared to a 2005 survey by Fidler, Burgman, Cumming, Buttrose, and Thomason (2006) of two prominent ecology journals, *Ecology* (EC) and the *Journal of Ecology* (JE) and two conservation biology journals, *Conservation Biology* and *Biological Conservation* (CB). For the three ETC surveys, 10 articles were randomly selected from each of the following journals: *Aquatic Toxicology, Archives of Environmental Contamination and Toxicology, Chemosphere, Ecotoxicology, Ecotoxicology and Environmental Safety, Environmental Pollution, Environmental Science and Technology, Environmental Toxicology and Chemistry, Marine Pollution Bulletin,* and *The Science of The Total Environment*. Features of each article were scored as Yes, No, or Not Applicable. Results are presented as frequencies and their associated 95% CIs. Most of the surveyed publications applied NHST (top panel) with all but a very few using a nil hypothesis of no

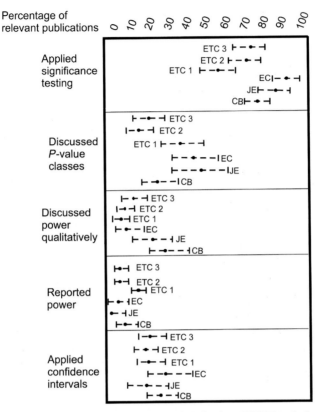

FIGURE 5.11   Survey documenting the continued misapplication of NHST methods.

difference, H_{nil}. A high proportion incorrectly discussed significance using categories, e.g., not significant, significant, or highly significant. However, there was evidence from the sequence of ETC surveys that the frequency of this category of P-value misinterpretation might be declining slightly. Few papers mentioned and even fewer reported statistical test power. About 10%–20% of publications applied CIs with no change through time being apparent from the ETC surveys. Clearly, more effort is required to improve inferences from statistical tests in our field.

The currently compromised NHST approach to making inferences from evidence is "a hybridization of Fisher's (1928) significance testing and Neyman & Pearson's (1928, 1933) hypothesis testing" (Anderson et al., 2000). Cohen (1994) applies the less tactful wording of a "mishmash of Fisher and Neyman-Pearson, with invalid Bayesian interpretation." Many of the separate components are sound but the NHST, as currently practiced, should be considered more a heuristic than a logically sound approach. For example, making inferences about alternative hypotheses during significance testing is illogical because no alternative hypothesis exists; however, the gross coincidence between NHST-based judgments and more logic-based judgments gives the current NHST practices the appearance of being "good enough." This seems an instance of a collective statistics heuristic in which the NHST results are grossly representative of those from logically well-founded methods. As discussed in Chapter 2, Human Reasoning: Everyday Heuristics and Foibles regarding heuristics, heuristics are often "fast and frugal," but they are more error prone than acceptable for scientific inquiry. Therefore, it is important to understand that they, as collective statistics heuristics, can lead to incorrect inferences. It is difficult to delineate the conditions in which they are good or bad inferential tools.

The failings of many statistical methods are becoming so obvious in many other scientific fields that a Kuhnian crisis is emerging. Attempts, such as the neoFisherian approach, adaptive inference, and wider use of CIs are being proposed to reduce the influence of the shortcomings of frequentist statistics. Alternative methods that avoid the shortcomings of current methods are coming into more frequent use due to recent innovations and the availability of computer software packages for their implementation. Chief among them are information-theoretic and Bayesian methods.

That statistical inferences are made in our field in a manner equivalent to a cognitive heuristic seems unacceptable to me. A transition from the conventional frequentist paradigm that arose in the early 20th century to a more reliable paradigm seems to be occurring. My personal opinion is that a series of stop-gap changes are required immediately and a combination of information-theoretic and Bayesian methods will emerge eventually as the new paradigm. This paradigm shift is well along in the statistical sciences. Sciences subject to careful scrutiny due to their subject matter, such as clinical or medical sciences, also seem to be moving away from the old paradigm. The warning against reporting NHST in the *British Medical Journal* (Altman et al., 2000; Fidler et al., 2004) is evidence of such change. Other sciences, such as ecotoxicology, are only slowly shifting away from the old paradigm. Perhaps the dual duty of many methods for scientific inference and legal regulation is particularly troublesome during attempts to shift toward a sounder paradigm. Standard methods embedded in regulations often are retained beyond their useful time for reasons other than sound scientific inference. Bacon's (1620) comment on idols of the theater (see Section 4.1.1) about pressures on workers within civil governments to oppose change seem relevant today.

Several recommendations might facilitate the soundest possible inferences during this time of transition.

- Avoid misinterpretation of NHST and CIs, moderating inferences so that they do not extend beyond the limits of sound interpretation.

- If applying NHST, attempt to define a meaningful ES and develop hypotheses other than the H_{nil}.
- Whenever possible, conduct *a priori* estimation of sample size to define α, β, and ES. This allows legitimate consideration of alternative hypotheses, estimation of PPV and NPV, and later Bayesian inferences.
- If possible and appropriate, apply the adaptive inference approach involving a sequence of tests with changing Type I and II error rates.
- If reasonable, use confidence or credible intervals that simultaneously provide insight about ES, precision, and if relevant, statistical significance.
- Report nonsignificant in addition to significant test results so that Bayesian analyses can be done with unbiased estimates of past studies that did or did not detect an effect, that is, produce an unbiased prior ratio or odds.
- Begin incorporating Bayesian or the information-theoretic means for making evidence-based inferences into your research efforts.

REFERENCES

Aitkin, M. (2010). *Statistical inference. An integrated Bayesian/likelihood approach*. Boca Raton: Chapman & Hall/CRC.

Altman, D., Machin, D., Bryant, T. N., & Gardner, M. J. (2000). *Statistics with confidence: Confidence intervals and statistical guidelines* (2nd ed.). London: British Medical Journal Books.

Anderson, D. R., Burnham, K. P., & Thompson, W. L. (2000). Null hypothesis testing: Problems, prevalence, and an alternative. *The Journal of Wildlife Management, 64*, 912–923.

Anderson, D. R., Link, W. A., Johnson, D. H., & Burnham, K. P. (2001). Suggestions for presenting the results of data analyses. *The Journal of Wildlife Management, 65*, 373–378.

Anderson, D. R. (2008). *Model based inference in the life sciences*. New York: Springer Science + Business Media, LLC.

Bakan, D. (1966). The test of significance in psychological research. *Psychology Bulletin, 66*, 423–437.

Bauer, H. H. (2012). *Dogmatism in science and medicine*. Jefferson, NC: McFarland & Co., Inc.

Bolstad, W. M. (2007). *Introduction to Bayesian statistics* (2nd ed.). Hoboken, NJ: John Wiley & Sons, Inc.

Borenstein, M., Hedges, L. V., Higgins, J. P. T., & Rothstein, H. R. (2009). *Introduction to meta-analysis*. Chichester: John Wiley & Sons, Ltd.

Bundschuh, M., Newman, M. C., Zubrod, J. P., Seitz, F., Rosenfeldt, R. R., & Schulz, R. (2013). Misuse of null hypothesis significance testing: Would estimation of positive and negative predictive values improve certainty of chemical risk assessment? *Environmental Science and Pollution Research, 20*, 7341–7347.

Bundschuh, M., Zubrod, J. P., Seitz, F., Newman, M. C., & Schulz, R. (2011). Mercury-contaminated sediments affect amphipod feeding. *Archive of Environmental Contamination and Toxicology, 60*, 437–443.

Burnham, K. P., & Anderson, D. R. (2001). Kullback-Leibler information as a basis for strong inference in ecological studies. *Wildlife Research, 28*, 111–119.

Carriger, J. F., Barron, M. G., & Newman, M. C. (2016). Using Bayesian networks to improve causal environmental assessments for evidence-based policy. *Environmental Science & Technology, 50,* 13195−13205.

Carver, R. P. (1978). The case against statistical significance testing. *Harvard Educational Review, 48,* 378−399.

Cohen, J. (1994). The earth is round (p < .05). *American Psychologist, 49,* 997−1003.

Cohen, J. (1995). The earth is round (p < .05). Rejoinder. *American Psychologist, 50,* 1104.

Cumming, G. (2012). *Understanding the new statistics. Effect sizes, confidence intervals, and meta-analysis.* New York: Routledge, Taylor & Francis Group.

Cumming, G., & Finch, S. (2005). Confidence intervals and how to read pictures of data. *American Psychologist, 60,* 170−180.

Di Stefano, J. (2004). A confidence interval approach to data analysis. *Forest Ecology and Management, 187,* 173−183.

Delignette-Muller, M. -L., Forfait, C., Billoir, E., & Charles, S. (2011). A new perspective on the Dunnett procedure: Filling the gap between NOEC/LOEC and ECx concepts. *Environmental Toxicology and Chemistry, 30,* 2888−2891.

Dixon, P. (2003). The p-value fallacy and how to avoid it. *Canadian Journal of Experimental Psychology, 57,* 189−202.

Doğan, I., & Doğan, N. (2017). The published research findings are trustable? Review. *Turkiye Klinikleri Journal of Biostatistics, 9,* 68−73.

Emerson, R. W. (2000). Essays: Second Series, Nature. In B. Atkinson (Ed.), *The essential writings of Ralph Waldo Emerson.* New York: The Modern Library. (Original work published 1844).

Fidler, F., Burgman, M. A., Cumming, G., Buttrose, R., & Thomason, N. (2006). Impact of criticism of null-hypothesis significance testing on statistical reporting practices in conservation biology. *Conservation Biology, 20,* 1539−1544.

Editors can lead researchers to confidence intervals, but can't make them think. In F. Fidler, N. Thomason, G. Cumming, S. Finch, & J. Leeman (Eds.), *Psychological Science* (15, pp. 119−126).

Fisher, R. A. (1925). *Statistical methods for research workers.* Edinburgh: Oliver & Boyd.

Fisher, R. A. (1958). The nature of probability. *Centennial Review, 2,* 261−274.

Gardner, M. J., & Altman, D. G. (1986). Confidence intervals rather than p values: Estimation rather than hypothesis testing. *British Medical Journal, 292,* 746−750.

Germano, J. D. (1999). Ecology, statistics, and the art of misdiagnosis: The need for a paradigm shift. *Environmental Reviews, 7,* 167−190.

Gigerenzer, G. (2000). *Adaptive thinking. Rationality in the real world.* Oxford: Oxford University Press.

Gigerenzer, G. (2004). Mindless statistics. *The Journal of Socio-Economics, 33,* 587−606.

Gigerenzer, G. (2008). *Rationality for mortals. How people cope with uncertainty.* Oxford: Oxford University Press.

Goodman, S. N. (1999). Toward evidence-based medical statistics. 1. The p value fallacy. *Annals of Internal Medicine, 130,* 995−1004.

Goodman, S. N. (2001). Of p-values and Bayes: A modest proposal. *Epidemiology, 12,* 295−297.

Guthery, F. S. (2008). Statistical ritual versus knowledge accrual in wildlife science. *Journal of Wildlife Management, 72,* 1872−1875.

Haller, H., & Krauss, S. (2002). Misinterpretations of significance: A problem students share with their teachers? *Methods of Psychological Research Online*, 7. Internet: http://www. mpr-online.de.

Harris, R. J. (1997). Reforming significance testing via three-valued logic. In L. L. Harlow, S. A. Mulaik, & J. H. Steiger (Eds.), *What if there were no significance tests?* Mahwah, NJ: Lawrence Erlbaum Associates.

Holling, C. S., & Allen, C. R. (2002). Adaptive inference for distinguishing credible from incredible patterns in nature. *Ecosystems*, 5, 319−328.

Howson, C., & Urbach, P. (1989). *Scientific reasoning: The Bayesian approach.* La Salle, IL: Open Court.

Hurlbert, S. H., & Lombardi, C. M. (2009). Final collapse of the Neyman-Pearson decision theoretic framework and rise of the neoFisherian. *Annales Zoologic Fennici*, 46, 311−349.

Hyndman, R. J. (1996). Computing and graphing highest density regions. *The American Statistician*, 50, 120−126.

Ioannidis, J. P. A. (2005). Why most published research findings are false. *PLoS Medicine*, 2, e124. doi:10.1371/journal.pmed.0020124.

Jaynes, E. T. (1976). Confidence intervals vs Bayesian intervals. In W. L. Harper, & C. A. Hooker (Eds.), *Foundations of probability theory, statistical inference, and statistical theories of science* (Vol. II, pp. 175−257). Dordrecht: D. Reidel Publishing Company Dordrecht-Holland.

Jeffreys, H. (1983). *Theory of probability* (3rd ed.). Oxford: Oxford University Press.

Johnson, J. B., & Omland, K. S. (2004). Model selection in ecology and evolution. *Trends in Ecology and Evolution*, 19, 101−108.

Kaiser, H. F. (1960). Directional statistical decisions. *Psychological Review*, 67, 160−167.

Kruschke, J. K. (2011). *Doing Bayesian data analysis. A tutorial with R and BUGS.* Burlington, MA: Elsevier Inc.

Kullback, S., & Leibler, R. A. (1951). On information and sufficiency. *Annals of Mathematical Statistics*, 22, 79−86.

Lehmann, E. L. (1950). Some principles of the theory of testing hypotheses. *Annals of Mathematical Statistics*, 21, 1−26.

Lesaffre, E., & Lawson, A. B. (2012). *Bayesian biostatistics.* Chichester: John Wiley & Sons, Ltd.

Morrison, D. E., & Henkel, R. E. (1970). *The significance test controversy.* New Brunswick, NJ: Transaction Publishers.

Mudge, J. F., Baker, L. F., Edge, C. B., & Houlahan, J. E. (2012). Setting an optimal α that minimizes errors in null hypothesis significance tests. *PLoS One*, 7, e32734.

Nakagawa, S., & Cuthill, I. C. (2007). Effect size, confidence interval and statistical significance: A practical guide for biologists. *Biological Reviews*, 82, 591−605.

Neapolitan, R. E. (2004). *Learning Bayesian networks.* Upper Saddle River, NJ: Pearson Prentice Hall.

Newman, M. C. (2008). "What exactly are you inferring?" A closer look at hypothesis testing. *Environmental Toxicology and Chemistry*, 27, 1013−1019.

Newman, M. C. (2013). *Quantitative ecotoxicology* (2nd ed.). Boca Raton, FL: Taylor & Francis/CRC Press.

Newman, M. C., Xu, X., Cotton, C. F., & Tom, K. R. (2011). High mercury concentrations in three deep ocean Chondrichthyans. *Archives of Environmental Contamination and Toxicology*, 60, 618−625.

Oakes, M. (1986). *Statistical inference: A commentary for the social and behavioral sciences.* Chichester: John Wiley & Sons, Ltd.

Pearl, J. (2000). *Causality. Models, reasoning, and inference.* Cambridge: Cambridge University Press.

Poole, C. (2001). Low p-values or narrow confidence intervals: Which are more durable? *Epidemiology, 12,* 291−294.

Popper, K. R. (1959). *The logic of scientific discovery.* London: Routledge.

Platt, J. R. (1964). Strong inference. *Science, 146,* 347−353.

Rizak, S. N., & Hrudey, S. E. (2006). Misinterpretation of drinking water quality monitoring data with implications for risk management. *Environmental Science & Technology, 40,* 5244−5250.

Rouder, J. N., Speckman, P. L., Sun, D.-C., & Morey, R. D. (2009). Bayesian t tests for accepting and rejecting the null hypothesis. *Psychonomic Bulletin & Review, 16,* 225−237.

Scutari, M., & Denis, J.-B. (2015). *Bayesian networks with examples in R.* Boca Raton: CRC Press/Taylor & Francis Group.

Simm, J., & Reid, N. (1999). Statistical inference by confidence intervals: Issues of interpretation and utilization. *Physical Therapy, 79,* 186−195.

Stigler, S. (2008). Fisher and the 5% level. *Chance, 21,* 12.

Sutton, A. J., & Abrams, K. R. (2001). Bayesian methods in meta-analysis and evidence synthesis. *Statistical Methods in Medical Research, 10,* 277−303.

Wacholder, S., Chanock, S., Garcia-Closas, M., El Ghormli, E., & Rothman, N. (2004). Assessing the probability that a positive report is false: An approach for molecular epidemiology studies. *Journal of the National Cancer Institute, 96,* 434−442.

Wagenmakers, E.-J. (2007). A *practical* solution to the pervasive problems of p values. *Psychonomic Bulletin & Review, 14,* 779−804.

Wasserstein, R. L., & Lazar, N. A. (2016). The ASA's statement on p-values: Context, process and purpose. *The American Statistician, 70,* 129−133.

Weber, C. I., Peltier, W. H., Norberg-King, T. J., Horning, W. B., Kessler, F. A., Menkedick, J. R., & Freyberg, R. W. (1989). *Short-term methods for estimating the chronic toxicity of effluents and receiving waters to freshwater organisms, EPA/600/4-89/001.* Cincinnati: US Environmental Protection Agency.

Woodworth, G. G. (2004). *Biostatistics. A Bayesian introduction.* Hoboken, NJ: John Wiley & Sons, Inc.

Ziliak, S. T., & McCloskey, D. N. (2007). *The cult of statistical significance.* Ann Arbor: The University of Michigan Press.

Section 3. How Groups Weigh and Apply Evidence

Chapter 6

Social Processing of Evidence: Commonplace Dynamics and Foibles

How could we have been so stupid?

John F. Kennedy as quoted in Janis (1971)

6.1 INTRODUCTION

Wishing never to repeat such a mistake, President John F. Kennedy asked the above question about his administration's 1961 failed attempt to overthrow Fidel Castro's government. He, with a small group of advisors, had secretly planned and overtly bungled the Bay of Pigs invasion. Kennedy initiated a review to determine what had gone wrong in the team's decision-making process, revealing group judgment errors that we now know to be common in advisory groups. This and other instances of failed decision-making prompted development of the concepts and partial remedies discussed in this chapter. The assumption here is that comparable decision-making errors can, and do, occur in ecotoxicology research and ecological risk assessments. An exploration might shed light on heretofore unexplained failures and perhaps reduce the risk of future errors.

Broadening Kennedy's question to include ecological risk assessment, the following questions might be posed. How could small, well-intended groups of respected professionals collectively come to the conclusion that environmental levels of atrazine were harmless despite mounting evidence to the contrary? (See Sections 2.2.2 and 2.3.1 of Chapter 2: Human Reasoning: Everyday Heuristics and Foibles, and Sections 4.1 and 4.2.2 of Chapter 4: Pathological Reasoning Within Sciences.) And why should past funding sources or previous membership in an ecorisk working group play such a strong role on this well-intentioned group's decision about the harmfulness of a chemical? On the other hand, what prompted the nonscientific affronts to this group of risk assessors by a self-appointed opponent? Labeling the associated process as pathological science as done in Chapter 4, Pathological Reasoning Within Sciences, does not answer these questions: it merely identified unintentional deviation from scientific norms. The reasons for these

The Nature and Use of Ecotoxicological Evidence. DOI: https://doi.org/10.1016/B978-0-12-809642-0.00006-6

departures and the means by which they might have been avoided remained unexplored. This chapter will examine the sociology giving rise to such suboptimal behaviors, suggesting ways of reducing the likelihood of their reoccurrence in ecotoxicology and ecological risk assessment.

Previous chapters focused narrowly on evidence generation, processing, and use by individuals. From this point onward, focus will be shifted to social cognition and co-decision-making by groups responsible for generating or applying evidence (Fig. 6.1). Group decision-making might involve a pair of individuals, small groups with or without formal structure, large networks or groups defined by some common feature or objective, or spontaneously coalescing collections of widely dispersed individuals like those that emerge on the Internet. It might also involve expert panels that combine individual, expert—elicitor pair, and group decision-making features.

This particular chapter deals with the social cognition and judgment involving pairs of individuals (dyads) and small groups, leaving features of larger groups and networks to later chapters. Understanding advice giving and receiving in dyads provides insight about ecotoxicology-related information processing in expert—decider arrangements. These dyads might include an anonymous reviewer giving advice to a journal editor about the acceptability of a manuscript, a researcher seeking the opinion of a peer about a particular experimental finding, or a graduate student seeking guidance from their advisor about how best to conduct their dissertation research. In some

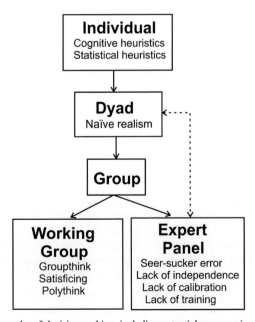

FIGURE 6.1 Hierarchy of decision-making, including potential compromises.

instances, a dyad is made up of two individuals with one making the final decision. More than one individual is involved but responsibility for the final decision or judgment rests in the hands of a single person, e.g., a journal editor deciding to reject a manuscript based on a reviewer's comments. In other instances, both parties might be peers who are equally responsible for a decision and must find a way to generate a consensus judgment. An example might be two co-investigators who jointly analyze experimental evidence with the intent of eventually publishing their mutual conclusions.

Decision-making by small groups is often distinct from that by dyads because the dynamics occur within a more complex social framework and responsibility might be spread across the entire group. Relevant to ecotoxicology and ecological risk assessment are small groups of co-investigators, advisory panels, risk assessment teams, and experts from whom opinions are being elicited. The members of some small groups might work cooperatively toward a decision or group members might be queried independently to minimize undue influence of some group members on the thinking of others. In some cases such as Delphi expert elicitation, the process is intentionally designed to blend elements of both, dyad and group decision-making. Independence also creates the opportunity to calibrate the effectiveness of each expert and to use the results to weight each expert's estimates before combining it with estimates of the other experts. In some elicitation designs, it also provides the opportunity for initial expert training (Kynn, 2008).

6.1.1   Evolution of Decision-Making Social Systems

Our innate social psychology is probably that bequeathed to us by our Pleistocene ancestors.

Richerson and Boyd (2005)

To this point, human history, judgment, and decision-making have been described in terms of how remarkable individuals are at thinking up increasingly efficient ways of exploiting environmental resources. This might have given the false impression that exceptional cognitive skills are the primary reason for our species' success. As suggested by the heuristics described in Chapter 2, Human Reasoning: Everyday Heuristics and Foibles, nothing could be further from the truth. Equally important is our sympatico, that is, our unparalleled capacity to understand one another and to use that understanding to achieve mutual goals. Social scientists who explore Theory of Mind (ToM) argue that sophisticated sociality, not intelligence, is the primary reason for our success.

... what seems special about humans is our desire and ability to understand what other people think and feel...this ability gives humans a remarkable power to make sense of other people's past behavior and to predict how their behavior will unfold given their past or future circumstances...Though there is

a conscious, reasoned component to ToM, much of our "theorizing" about what others think and feel occurs subliminally, accomplished through the quick and automatic processes of our unconscious mind.

Mlodinow (2012)

Just as individual cognition is a blend of subconscious heuristics and deliberate reasoning, human sympatico involves instinctive shortcuts and well-reasoned predictions about the thoughts of others. Like cognitive heuristics, social shortcuts can permeate group scientific judgments and risk assessments without our conscious assent, potentially degrading the quality of mutual conclusions. As an important example, social validity can inadvertently trump scientific validity because stability of social processes requires a certain degree of consistency. "... the goal of the [group] decision process is to see the world with confidence rather than accuracy... people seem to seek not certainty of knowledge but social validity" (March, 1994). Without deliberate application of safeguards, social validity and cohesion can become more important than physical or scientific validity of judgments (March, 1994; Miller, Garnier, Hartnett, & Couzin, 2013).[1]

There is no reason to believe that groups of ecotoxicologists and ecological risk assessors are any less prone to this behavior than other groups. Most authentic ecological risk assessors would grudgingly agree that some customary steps and methods are applied more to conform to expected group norms than to enhance scientific soundness. In my opinion, the common practice of deriving a reference dose by dividing no observed effect level (NOEL) values by uncertainty and modifying factors (often 10) is one example in which social validity and cohesion trump toxicological validity.

The ToM concept of intentionality provides a glimpse into the complexity and inherent risk of misinterpretation during social reasoning (Mlodinow, 2012). Humans are unique in that they are capable of formulating very complex theories about the mental state of others, that is, forming theories of mind and making predictions from those theories. Individuals anticipate and modify their own behavior based on surmises about other's beliefs, intents, trustworthiness, possible reactions, and emotional states.

ToM intentionality refers to how many steps are being considered away from an individual's own state of mind when predicting state of mind. A first-order intentional refers to an individual's ability to contemplate their

---

1. Exceptions can be found to this sweeping generality. An admirable example is the scientific fortitude displayed by Baas Kooijman who wrote in his paper introducing the formalized SSD technique, "All honest scientific research workers will feel rather uncomfortable with such a task [of helping establish the SSD method], and the author is no exception. The feeling finds its roots in the extrapolation of experimental findings far beyond the limits of our knowledge" (Kooijman, 1987). This statement certainly did not enhance the social stability for the group trying to integrate the method into their activities.

own state of mind. Many nonhuman species are capable of first-order intentionality. For a human ecotoxicologists or risk assessor, a first-order intentional might be reflected in a statement like, "Based on the evidence, I believe that the first explanation is most credible." A second-order intentional involves pondering the state of mind of someone else. For an ecotoxicologists, a second-order intentional might be indicated in the statement, "After she reviews the evidence, I think Dr. Yeats will favor the first explanation too." Some nonhuman primates such as chimpanzee seem capable of limited inferences involving the second-order intentional (Mlodinow, 2012; Premack & Woodruff, 1978). A third-order intentional extends speculation one step further to include another person's thinking, such as, "Dr. Yeats will feel that Dr Watts will be disappointed when she states that the first explanation is the most credible one." A higher-order intentional might add still another mind, as in, "I think that Dr Yeats will be concerned that the committee's final report drafted by Dr Watts might suggest to the sponsoring government agency head that she (Yeats) and I were equivocal about the first explanation's credibility." Only humans are capable of intentionals extending on occasion as high as sixth-order.[2] If the reader tries to imagine for a moment the large number of various order guesses occurring in a small group of humans judging the risk of harm from a contaminant, it is easy to see why "social heuristics" need to be thoughtfully managed to avoid confusion and error.

In the context of intentionality, the first half of this book described mostly aspects of first- and occasionally second-order intentionals that put at risk an individual's ability to make optimal decisions. This chapter extends decision-making to that involving dyads and small groups. Intentionality most pertinent to dyads is first and second order. Examples might be the statements, "I judge the risk to be unacceptable" or "My co-investigator will come to the wrong conclusion about the risk associated with this contaminant. So I need to develop a good defense of my conclusion of unacceptable risk." Intentionalities extending to fourth-order are common for small groups. As a high-order example of an intentionality in a small group is the guess, "I conclude that Dr Jonas is biased, and the panel chair will realize this and downplay her impact in the final report without my having to say something." The risk of misjudgment multiplies as intentionality order increases. If left unchecked in a group, it is likely that these errors will lessen decision quality. Chapters following this one will describe the nature and potential errors in decision-making associated with miscomprehending higher order intentionals.

---

2. As an aside, the compromised capacity to form theories of mind is characteristic of schizophrenia (Sprong, Schothorst, Vos, Hox, & Van Engeland, 2007) and the autism spectrum (Baron-Cohen, Leslie, & Frith, 1985).

6.1.2   Homophily and Heterophily

As already discussed, an overarching theme in human evolution is enhanced fitness by maximizing cooperation within small groups of related individuals and minimizing conflict among distantly related groups that are competing for shared resources. Relatedness for a Pleistocene hunter–gatherer group largely meant genetic kinship but it encompasses more cultural affiliations for modern groups.[3] Similarly, what is considered shared resources has expanded beyond those needed by Pleistocene hunter–gatherers to kept body and soul together. Key resources now include access to decision makers, monetary resources, knowledge, power, and prestige.

The nature of several social tendencies becomes important to understand at this point and the complementary concepts of homophily and heterophily provide a good starting place for doing so. Homophily is the human tendency to interact, associate, or form stable relationships with individuals who share beliefs, background, or other characteristics. Simply put, humans are more likely to form ties with individuals like themselves than with those unlike themselves. During our Pleistocene evolution, a homophilic group would have been a small, kinship-based group. European Union-affiliated university professors who study neonicotinoid ecotoxicology might be a homophilic subgroup within a large advisory panel studying the global decline of crop pollinators. Beliefs, available information, and potential biases would be more homogeneous within this subgroup than in the panel as a whole. The advantage of homophily in a group is the reduction of interpersonal dissonance. It promotes group strategic action, harmony, trust, and loyalty (March, 1994) because the intentionalities of members are easily predicted by other members.

Homophily within our field can be driven by many factors such as common professional standing, highest earned academic degree, professional role, regulatory framework in which one works, organizational affiliation, first language, core values, position in a particular social network, or even research vantage (reductionistic or holistic) (McPherson, Smith-Lovin, & Cook, 2001). Some contributors to homophily are a matter of choice (e.g., an individual's core values or research vantage), while others are imposed (e.g., an individual's geographical location or regulatory milieu) (McPherson & Smith-Lovin, 1987). Choice-based homophily results from an individual's preferences, and accordingly, is also called individualistic homophily. In contrast, contributors to homophily such the regulated industry in which

---

3. As briefly described in Section 1.4 in Chapter 1, The Emerging Importance of Pollution, one can envision genetic selection being supplemented, or perhaps supplanted, by meme selection in this instance. Memes are units of cultural inheritance that are learned, practiced, and potentially transmitted to others. A meme is favored if it enhances cooperation within groups and minimizes conflict among groups.

one finds oneself employed is an induced or structural factor. Geographic distance is an important structural factor influencing homophily although its influence has been weakened slightly by internet technologies. Together, choice- and structure-based mechanisms give rise to homophily in ecotoxicology and risk assessor groups. Homophily-driven dyadic ties can extend outward through friend-of-a-friend ties, resulting in triad (groups of three individuals) and more complex patterns of connectivity (Knoke & Yang, 2008). Relative to social network discussions later in Chapter 8, Evidence in Social Networks, homophily can influence the large networks in which ecotoxicologists and risk assessors find themselves.

In contrast to homophilic groups, heterophilic groups are comprised of individuals with different sets of beliefs, backgrounds, biases, and other features. For the Pleistocene hunter—gatherers, heterophilic interactions occurred among small, genetically distant clans occupying overlapping territories. Social evolution gave rise to human behaviors that minimize disruptive conflicts among heterophilic groups competing for shared resources. As a modern example, a heterophilic collection in a large advisory panel considering pollinator decline might include the members from regulatory agencies, representatives of regulated industries, agriculturists, epidemiologists, academics, and employees of concerned NGOs from different regions of the world. Another set of heterophilic groups in this large advisory panel might be groups operating under different sets of environmental laws and regulations. Not surprisingly, dynamics and common decision-making errors are different for homophilic and heterophilic groups.

The purpose of bringing up the concepts of homophily and heterophily is that they can create conditions for both enhanced and degraded judgments. Two examples can illustrate this point and be interpreted in the context of ecotoxicology and risk assessment.

Rauwolf, Mitchell, and Bryson (2015) examined the influence of homophily on group cooperation and willingness to rely on incorrect social information (i.e., false gossip). Unsurprisingly, they found that homophily has the beneficial influence of enhancing group cooperation. However, this enhanced cooperation came at a price. They also found that homophily enhanced a group's willingness to accept error-prone, third-party gossip. So, homophily can degrade the soundness of information shared within a group. It is not too difficult to extend these findings to predict that groups of homophilic ecotoxicologists or ecological risk assessors will be more prone to use poorly substantiated evidence than heterophilic groups.

Assuming that the spread of sound, novel information within social media groups such as contributors to a blog was influenced by homophilic and heterophilic group members, Yamamoto and Matsumura (2009) attempted to predict the optimum blend of members that resulted in the most effective word-of-mouth communication. On the one hand, if the exchange occurs only between homophilic individuals, an information sender is unlikely to know more than the individual receiving their communication. On the other hand, in an exchange between two very heterophilic individuals, the unfamiliarity of the receiver with the material conveyed by the sender might cause enough dissonance to inhibit effective communication. As a predictable complication, exchanges between homophilic individuals are easier and consequently will occur much more frequently than those between heterophilic individuals. Yamamoto and Matsumura (2009) found that the most effective communicators of information (elites) were characterized by a specific blend of homophily and heterophily relative to the pool of information receivers. These findings about the influence of homophily and heterophily on the most effective diffusion of new information into a group are relevant to ecotoxicologists. An ecotoxicology student who understands only the information common to their homophilic group is very unlikely to become an information elite. A student wishing to increase their understanding of pollutant fate and effects in the biosphere is well advised to master established facts in ecotoxicology but also to seek information and concepts from disciples outside their immediate field. The student's eventual communication of unfamiliar information into ecotoxicology should involve its skillful commingling with information that ecotoxicologists already know.

6.2 INDIVIDUAL—PEER DYADS

Decision-making and estimation by dyads primarily involves consideration of first- and second-order intentionalities. The dyads that will be emphasized here are those involving advice taking and mutual decision-making.

> Common sense is the most widely shared commodity in the world, for every man is convinced that he is well supplied with it."
>
> Descarte (1637)

There is a well-worn quip that goes, "Most people think that what the world needs is more commonsense — but they themselves have more than enough." Most people are too confident of their own cognitive powers. In Chapter 2, Human Reasoning: Everyday Heuristics and Foibles, this notion was given the formal label of overconfidence bias. The relevance of this bias in environmental science can be underscored with my October 2016 survey conducted at a joint Association of Ecosystem Research Centers and the International Union of Radioecology (AERC-IUR) Workshop (Savannah River Site in South Carolina, USA). The survey asked for audience members to answer the opening questions of Chapter 2 about the likelihood of a tap water sample containing a detectable concentration of pesticide and Chapter 5 about the interpretation of P-values. However, they were also asked to indicate how confident they were with their answers (Table 6.1).

TABLE 6.1 Confidence in Correct and Incorrect Answer to Opening Questions of Chapters 2 and 5[a]

Question	Answer Correctness	Mean Confidence[b] (N, 95% HDI)
Chapter 2	Correct	0.807 (7, 0.584−1.021)
	Incorrect	0.693 (18, 0.596−0.787)
Chapter 5	Correct	0.903 (11, 0.831−0.975)
	Incorrect	0.788 (13, 0.675−0.897)

[a]Recollect that the question from Chapter 2 involved judging the likelihood of pesticide presence in tap water and that from Chapter 5 involved picking the correct definition of a P-value. Note that these questions had been asked before at the Savannah River Ecology Laboratory, so several of the responders had likely taken the survey previously. In terms used earlier, they had considerable "local" knowledge. So the range of confidence in answers was wider than would be expected for a naïve set of responders. The frequency of correct answers was also higher. These factors created ideal conditions for illustrating overconfidence.
[b]The scale used for confidence was the following probability of being correct: 1.00, 0.90, ..., 0.10 or less.

Although those picking the correct answers were more confident than those picking the incorrect answers, overconfidence was apparent in incorrect choosers of both questions. In the case of opening question of Chapter 2, there was no simple trend in the level of overconfidence as a function of how far the answer deviated from the correct answer.

Overconfidence bias of individuals bleeds over into how they combine their own judgments with those of others. Social scientists refer to this innate bias when comparing one's opinion with that of another as naïve realism, that is, the "conviction that we see matters 'objectively' and that insofar as others disagree, it is due to error or 'bias'[on their part]..." (Liberman, Minson, Bryan, & Ross, 2012). Naïve realism creates a challenge for dyads wishing to make the best joint judgment.

Naïve realism impedes optimum dyad use of estimates or judgments for several reasons. Take the situation in which both members of the dyad have equal backgrounds and abilities, but only one is responsible for making the final judgment. An example of this type of dyad might be a researcher informally discussing a new experimental finding with a colleague. Another might be a team leader discussing ecological effects data with a fellow risk assessor. During the examination of evidence, the decider formulates their own tentative answer and then receives that of their peer. Several approaches can be used to combine the two answers including unweighted and weighted averages. The most typical approach to accommodating peer input is to weigh one's own answer more heavily than that of the other. As discussed in terms of overconfidence bias in Chapter 2, Human Reasoning: Everyday Heuristics and Foibles, Yaniv (2004) found a general weighting of 70% for one's own opinion and only 30% for that of the advisor.

Naïve realism and heavily weighting one's own judgment results from several factors, not simply vanity. A certain degree of anchoring is inevitable because one forms their own opinion and then that of the other person is added to it (Liberman et al., 2012; Tversky & Kahneman, 1974). The details and judgments behind one's own opinion are more easily recalled than those of the other person. There is also some second-order intentionality guesswork required to process an advisor's opinion (Yaniv, 2004). The net result of anchoring, easier recollection of the grounds for one's own judgment, and uncertain intentionality result in overweighting of one's own judgment.

Various factors can influence the magnitude of weighting. Perceived superior experience of the advice giver and degree of dyad homophily lessen underweighting of an advisor's opinion (Bonaccio & Dalal, 2006; Harvey & Fischer, 1997). Despite the poor correlation between an individual's confidence and judgment, deciders might also unwisely give more weight to an advisor's opinion if the advisor seems very confident (Bonaccio & Dalal, 2006). Price and Stone (2004) refer to this tendency in advice taking as the confidence heuristic. Likely, weighting is also influenced by the fluency heuristic described in Section 2.3.2.7 in Chapter 2: underweighting will be less extreme if the advisor's opinion is expressed in a clear and fluent manner versus an obtuse and complicated way.

As the importance of a decision increases, there is a tendency for deciders to increase the weight they give to the other's opinion (Harvey & Fischer, 1997). Bonaccio and Dalal (2006) suggest this increased openness is prompted as much by a desire to share accountability for the decision as by any desire to improve the decision. The overweighting of a decider's own answer might also increase as the discrepancy widens between the estimates of the decider and advisor (Liberman et al., 2012). This includes the extreme case of the advisor's input being completely ignored if it is too divergent from the decider's own judgment (Harvey & Fischer, 1997). Such cognitive trimming (Yaniv, 1997) of extreme opinions will be discussed again in dealing with outlying opinions in groups.

In the long run, unweighted averaging of inputs of the decider and the advisor provides better decisions than the underweighting of the advisor's input just described (Yaniv, 1997). Nonetheless, it is a common practice for deciders to give unequal consideration to their own and a credible advisor's advice (Bonaccio & Dalal, 2006). Even when two judgments are presented and discussed openly by the decider and advisor, deciders tend to adjust their initial judgments by only a token amount (Bonaccio & Dalal, 2006).

Several approaches to advice giving and taking exist, and the influences of the above factors are different for each. First, in the long run and all else being equal, unweighted averaging of the two judgments is the most reliable strategy. Unfortunately, the natural temptation to do otherwise creates enough discomfort for many deciders that they invoke *ad hoc* rationales for deviating from this approach. For example, they might decide that enough doubt exists about the "in the long run and all else being equal" condition that weighting is required. Or the decider might reason that their opinion should carry more weight because they will be held more accountable for a bad decision than the notionally less informed advisor.

Bonaccio and Dalal (2006) assessed three schemes for generating a numerical answer from a dyad. In one experiment, the dyad members were required to present their initial estimates and then discuss the reasoning behind them before finalizing their estimates. Although individuals only adjusted their final answers by approximately 30% in the direction of the other dyad member's estimate, including discussion and consequent adjustment of answers did improve estimate accuracy relative to estimates from dyads that had no controls on naïve realism. The more divergent the initial estimates, the less the members adjusted their estimates in the direction of their partner. The next approach involved a bidding process in which the pair exchanged written bids without discussion until the process converged on a single number. This approach was less effective than the discussion approach in improving accuracy although it did improve accuracy relative to decision-making with uncontrolled naïve realism. The improvement was no better than would have been obtained by taking the unweighted average for the initial estimates.

The conclusions from Bonaccio and Dalal (2006) provide useful guidance. It was apparent that discussion with the intent of possibly adjusting one's initial estimate improved decision accuracy of a dyad. When initial estimates were very divergent—a condition in which simple averaging is the most advantageous—unconscious processes result in individuals being resistant to adjusting their estimates in the direction of the other dyad member. Unfortunately, highly divergent estimates did not prompt members to question and more carefully review their own initial estimates. Members also did not seem willing to consider the possibility that the initial answers were divergent because they were based on different, but equally sound information. In such a situation, decision accuracy can be improved by dyads engaging in back-and-forth discussions.

The improvement to a decider's estimate produced by integrating the opinion of another brings up the question about whether adding still more opinions or estimates will produce an even better final estimate or judgment. If that is the case, how many peers should one engage to have the best chance of minimizing subjective bias and drawing on as much knowledge as practical? In the situation in which advisors' inputs are independent of one another, the surprising answer is not many at all. Yaniv (2004) suggests "...a small number of opinions (e.g., three to six) is typically sufficient to realize most of the accuracy gains obtainable by aggregation." This issue will be examined again in Section 6.5.

6.3   GROUPTHINK AND POLYTHINK

Can a group or team[4] produce a better answer than an individual or simple dyad? The answer is yes under many conditions but no under others. Groups consistently outperform individuals when addressing a problem that has a clearly correct answer, and most often, does so by adopting the opinion of the most capable group member (Kämmer, Gaissmaier, & Czienskowski, 2013). Groups also provide superior results under other schemes such as one that carefully aggregates all members' opinions. However, an unthoughtful aggregation scheme can produce defective decisions.

Opinion aggregation to generate a group decision might involve weighting as mentioned above for dyads. Informal weighting can produce bias if strongly influenced by features such as perceived homophily/heterophily of group members. Thoughtful weighting as illustrated in Section 6.5 can reduce the risk of a degraded decision. With working groups, attempts to improve a final judgment might also involve trimming of extreme estimates or opinions (Yaniv, 1997). Successful trimming requires much care because outlying estimates or opinions might actually contain valuable information unknown to other group members. If done properly, decision accuracy can also be improved by weighting after trimming outlier opinions.

> *[Groupthink refers] to the mode of thinking that persons engage in when concurrence-seeking becomes so dominant in a cohesive ingroup that it tends to override realistic appraisal of alternative courses of action ... the term refers to a deterioration in mental efficiency, reality testing and moral judgments as a result of group pressures.*
>
> Janis (1971)

The decision to invade Cuba that so tormented President Kennedy was a prototypical instance of the group error called groupthink (Janis & Mann, 1977). Janis (1971) intentionally gave this error a name with an Orwellian timbre.[5] He did so to underscore groupthink's ability to create a distorted group reality. The overriding theme in groupthink is "loyalty requires each member to

---

4. The qualities of work groups or teams as outlined by Kozlowski and Ilgen (2006) are assumed in this chapter with one minor difference. According to Kozloski & Ilgen, a working group is made up of two or more interdependent members who work together to accomplish some task(s) relevant to a larger organization, e.g., a group providing a report to a responsible regulatory agency about a new pesticide's potential risk. Group members can take on different roles, e.g., leader or primary report author. The trivial change made in this chapter is to increase the group size slightly to three or more individuals. It was useful in this book to discuss dyads separately because they play unique, key roles in ecotoxicology and risk assessment activities that triads and larger groups do not.
5. In George Orwell's classic novel, *Nineteen Eighty-Four*, a dystopic society was controlled partly with a form of brainwashing called doublethink. Doublethink, the ability to accept two logically inconsistent beliefs or thoughts simultaneously, served to reinforce illogical political party beliefs. Janis's groupthink was an intentional twist on this term.

avoid raising controversial issues" (Janis, 1982). The central goal is alleviation of predecisional anxiety by quickly reaching an ostensible consensus, instead of an optimal decision reached by deliberate and potentially exhausting deliberation. Relative to scientific judgment, groupthink seems the antithesis of Chamberlin's method of multiple working hypotheses (see Section 3.2 in Chapter 3: Human Reasoning: Within Scientific Traditions and Rules).

The groupthink concept was illustrated originally with well-known events that immediately caught the general public's attention (Turner & Pratkanis, 1998). Janis's examples were directly pertinent to people's lives during the 1970s and explained previously enigmatic political decisions. Furthermore, Janis's articles and books presented the concept with exceptional fluency (see Section 2.3.2.9 in Chapter 2: Human Reasoning: Everyday Heuristics and Foibles, "fluency heuristic"). Understandably, the concept diffused quickly throughout the social sciences and then into the public vernacular (Fig. 6.2). The concept continued to be refined in the social sciences but remained as initially described in the public's mind. Its outdated conceptualization by many and continued impact on flawed decision-making necessitate detailed discussion of it here.

Irving Janis's original description laid out preconditions for groupthink, including the existence of a provocative situational context, little hope of a better decision than that advocated by the group leader, perceived high stress exerted from some outside source(s), and temporary doubt perhaps due to a recent failure, the difficulty of the task, or a moral dilemma (Janis, 1971; McCauley, 1998). Groups experiencing groupthink tended to be uncritical of their leader, a behavior most readily adopted by a group

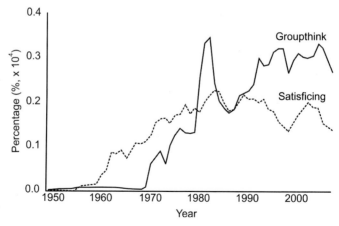

FIGURE 6.2 N-Gram analysis of the terms, groupthink (solid line) and satisficing (dashed line) from 1950 to 2008. The y-axis indicates the frequency (as a percentage) of the term's occurrence in all Google-scanned English books. (see https://books.google.com/ngrams/info, for more details about N-Gram.).

with strong directive leadership (Janis, 1971; Janis & Mann, 1977).[6] Although these preconditions were apparent in the original situations portrayed by Janis, he formulated the initial groupthink concept based on failed decisions with calamitous consequences. Key examples include the failed Bay of Pigs invasion, failure of US Pearl Harbor forces to prepare for the Japanese invasion, the Johnson administration's misjudging of the consequences of escalated bombing of North Vietnam, and later, the Challenger space shuttle disaster. By their nature, the situations were likely to have several of these preconditions, such as the existence of a provocative situation and substantial stress from external sources.

Studies conducted in the decades since the introduction of groupthink found that, although these preconditions do foster groupthink, they are not necessary for it to emerge. McCauley (1998) suggests that groupthink can emerge whenever discomfort induced by high uncertainty makes a premature consensus attractive. Janis (1982) also found during his analysis of the Watergate break-in fiasco that homogeneity of members' ideology (value homophily) was a strong promoter of groupthink. Later, he (Janis, 1989) modified his original list of preconditions, indicating that susceptible groups "are [often] highly cohesive, insulated from experts, perform limited search and appraisal of information, operate under directed leadership, and experience conditions of high stress with low self-esteem and little hope of finding a better solution to a pressing problem than that favored by the leader or influential members" (Turner & Pratkanis, 1998). Baron (2005) broadened these preconditions even further, arguing that "concurrence-seeking, illusion of consensus, self-censorship, and ingroup defensiveness ... are far more widespread phenomena than [Janis] envisioned." Groupthink occurs in many commonplace activities other than the original, catastrophic ones portrayed by Janis.

As a relevant example, Moorehead, Neck, and West (1998) and Neck and Manz (1994) argue that they expect the risk of groupthink to increase as organizational management frameworks continue to move toward self-managing teams (SMTs)[7] that is, "groups of 4 to 12 individuals who share responsibility for completing relatively whole tasks" (Moorehead et al., 1998). Self-managing teams are given discrete tasks and responsibility for autonomously generating a sound decision or product. They have substantial decision-making powers with the role of managers shifting from final decision maker to that of facilitator. Moorehead et al. (1998) observed that, if

---

6. A directive leadership style is one in which a single person establishes the direction of the team and makes all final decisions. It is a tight-rein leadership style as opposed to a loose-rein style in which the leader gives considerable leeway to a highly motivated and well-trained team who complete the task together as equal partners.

7. Qualities of SMTs that relate directly to preconditions for groupthink are summarized in table 1 of Moorehead et al. (1998).

orchestrated unthoughtfully, self-contained SMTs can become isolated from experts in the larger organization who might have otherwise enhanced their decision or product quality. Although SMTs have considerable decision-making powers, members often lack sufficient management experience to avoid errors such as groupthink. Norms for decision-making in the SMT can be ill defined or nonexistent. Interpersonal cohesion can also be higher within than between SMTs, creating the risk of undue influence of homophily. The pressure on an SMT to produce a product or decision in a timely fashion can create a source of external stress and also a temporary loss of member self-esteem if the SMT has failed to deliver a sound decision or product in its recent past. Moorehead et al. (1998) argued that some level of managerial training is needed for SMT members to avoid groupthink and other decision-making missteps.

Self-managing teams are commonplace in ecotoxicology and environmental risk assessment although they are not called SMTs. Examples include research panels assigned by a scientific organization to produce a white paper on an emerging issue or groups developing risk assessment reports for a Superfund site operable unit. They emerge naturally in the collegial style of decision-making widely practiced in academia, providing fertile ground for groupthink (Klein & Stern, 2009). If insufficient attention is being paid, there is the danger of academics working in ecotoxicology scientific teams or regulatory panels transplanting bad habits learned at their home institutions into these activities.

Table 6.2 summarizes the characteristics of a group falling victim to groupthink. All are linked to the group's subconscious desire to reduce pre-decisional stress by working more toward group concurrence than an optimal decision. The members wrongly emphasize or exaggerate group invulnerability, superiority to outgroups, superior morality, and unanimity. In ecotoxicology and environmental risk assessment, superior morality might involve a feeling of superior insight or experience needed to move society toward real-world solutions to pollution. Pressure on dissenters in the group, tacit support of a consensus by not voicing one's doubts, and self-censoring take many forms.

In my experience, groupthink is common in ecotoxicology and risk assessment working groups or panels. It is commonplace to hear during one-on-one conversations with group members that they have strong doubts about the group's emerging decision but that they remain silent because they wish to be good team players.[8] Outgroup stereotyping is also quite common.

---

8. Of course, it is extremely difficult during these exchanges to determine whether the member expressing doubt is being truthful about their state of mind or simply agreeing politely with someone who they suspect has reservations about the group decision. The uncertainty characteristic of intentionality prediction makes it very difficult to choose confidently between these alternatives.

TABLE 6.2 Symptoms of Groupthink Outlined in Janis and Mann (1977)

1. An illusion of group invulnerability emerges and leads to overconfidence and heightened risk taking.
2. Collective rationalization is applied to reduce the importance of any contrary facts or counterarguments, allowing the group to remain committed to the already favored decision.
3. Outgroup stereotyping or dehumanization appears that permits the group to proceed without considering outgroup opinion and evidence, or seeking dialogue with them.
4. Belief in the inherent superior morality of the group (and inferior morality of outgroups) that results in moral objections being left unsaid. Eventually the group creates a sense of superiority that can free them to make what would normally be considered moral transgressions. In ecotoxicology, a "moral transgression" would include adjusting belief based on illogical thinking, unsound science, or contrary to in-hand evidence.
5. Pressure is applied to dissenters in the group who try to bring up counter evidence, allowing the group's shared decision framework and decision to move ahead without additional stress.
6. Group members who might object display self-censoring in order to foster and maintain group cohesiveness.
7. An illusion of unanimity in the group is created and maintained by self-censoring and other mechanisms.
8. Self-appointed mindguards emerge who take on the role of protecting the group from contrary information that might lessen the group commitment to and level of comfort with the emerging consensus decision.

A member of an outgroup proposing an alternative decision might be dismissed as being naïve, "a fuzzy-headed, ivory tower academician incapable of working toward practical solutions to real world problems," "a paid consultant who is only there to protect their client's interests," or "a regulatory hack who has forgotten—or never understood—basic ecological realities." That individual is often penalized (Baron, 2005), perhaps by not being invited onto future panels dealing with similar decisions. Mindguards emerge as needed (see Table 6.2, Point 8). I recollect working on an EPA panel and proposing a second option to complement a favored one that had no ecological justification. I mentioned a demographic approach that would provide insights about population-level effects. After mentioning this option twice during the proceedings, one member who had to that point contributed little to the panel deliberations expressed frustration at my repeatedly mentioning demographic methods without giving enough detail and justification for such a novel approach. His *ad hoc* objection allowed the group to turn its attention back to the favored option. An important point is that his successful objection was absurd. Demographic methods were established in the 1660s during the second wave of plague in Europe and now enjoy widespread acceptance in other life sciences. They are also becoming increasingly commonplace tools in ecotoxicological studies. There was nothing novel or untested about them.

  Some consequences of groupthink, as outlined by Janis (1971) are the following:

FIRST, the group limits its discussions to a few alternative courses of action (often only two) without an initial survey of all the alternatives that might be worthy of consideration.

SECOND, the group fails to reexamine the course of action initially preferred by the majority after they learn of risks and drawbacks they had not considered originally.

THIRD, the members spend little or no time discussing whether there are non-obvious gains they might have overlooked or ways of reducing the seemingly prohibitive costs that made rejected alternatives appear undesirable to them.

FOURTH, members make little or no attempt to obtain information from experts ... who might be able to supply more precise estimates of potential losses and gains.

FIFTH, members show positive interest in facts and opinions that support their preferred policy; they tend to ignore facts and opinions that do not.

All of these consequences were apparent in the EPA panel mentioned above in which a mindguard emerged. This panel functioned as an SMT responsible for one report that would be integrated by the organizers with those of three other panels to produce a final report. Relative to the first groupthink consequence, the group (including myself) quickly focused on one easily-implemented option that notionally predicted threshold pesticide environmental concentrations that could have unacceptable effects in aquatic ecosystems. The approach had been successfully advocated in other recent working groups that included individuals who were now also members of this new EPA panel. The shortcomings of the advocated approach (the second and fifth consequences) and advantages of alternative approaches (the third consequence) were explored briefly and the advocated method deemed "good enough" despite its drawbacks. In reality, the advocated method used a small set of laboratory-derived LC50, EC50, NOEC, and LOEC metrics of effect to individual organisms to imply harmfulness to entire ecological communities. In contrast to Kooijman's discomfort in extrapolating beyond the limits of existing ecotoxicological knowledge (see Section 6.1.1), the panel seemed quite comfortable judging their approach to be adequate for the task. Relative to the fourth consequence, after an initial attempt to thoughtfully discuss other options was met with indifference by the panel leader, no panel members objected again as the decision-making process progressed. My early attempt to offer an alternative prompted the mindguard response just described. Periodically, a fellow panel member would privately express simultaneously their doubts about the approach and their decision to act collegially in this situation. The approach eventually became central in the final report, and a tee shirt with a figure illustrating the new approach on it was presented to each member at the end of the panel's work. In this way, the mandate to incorporate "the latest scientific knowledge and best current technology" into activities was somehow satisfied.

Polythink, also named along the lines of Orwell's "doublethink," is another dysfunction in groups that results in no sound decision from heterophilic teams. Positioned at the opposite extreme to groupthink on the continuum of possible group behaviors, polythink occurs if "a plurality of opinions and views leads to disagreement between group members" (Mintz & Wayne, 2016). So much heterophily exists that agreement on a single, optimal decision becomes impossible. Strom, Stansbry, and Porter (1995) gave the example of radiation protection decisions being subject to polythink because the involved regulators, health physicists, other scientists, and (defendant and plaintiff) lawyers worked under very conflicting paradigms. (See value centrality bias in Chapter 2.) Distinct and intractable stances on whether the relationship between received radiation dose and adverse effect was linear at low doses were taken based on the professional premises of a particular group member. A regulator drew on agency publications to reinforce their opinion. A scientist judged that a linear hypothesis is the hypothesis with more support than a threshold hypothesis. A lawyer for a plaintiff would argue for a linear model that predicts a risk of harm at even low very doses.

If decision-making is taking place with a deadline or other constraints, leadership intervention might be required to bring the group suffering from polythink to a conclusion, often resulting in a suboptimal decision (Mintz & Wayne, 2016). An example of a group working under a time constraint might be an ecological risk assessment team that is legally or contractually required to report their conclusions by a certain deadline. Polythink in that situation would be a formidable challenge. Often these groups and groups dealing with crises will have strong leadership to avert or minimize polythink (Baekkeskov, 2016).

A moment of reflection makes it clear that a level of polythink is present at certain times in any healthy science, even the applied science of ecotoxicology. Concepts and paradigms can only evolve if enough contrasting opinions based on different evidence emerge to cause a paradigm crisis. However, the level of polythink in a scientific context must not result in incomplete consideration or premature discarding of alternatives. From the vantage of this book, polythink is a dysfunctional process under certain conditions but a necessary one under others. Whether intervention by leadership is wise when a group succumbs to polythink depends on whether the group is attempting a scientific or a constrained regulatory judgment. Even in a regulatory process, polythink can have some beneficial features. Mintz and

Wayne (2016) make the following point about the benefits of polythink in a nonscience context *if managed effectively.*

> *... in many cases, political officials/business executives/managers/other group leaders can benefit from polythink if they leverage positively the plurality of opinions presented to them in a group setting en route to forming a decision. For example, if a leader can take the diverse feedback of group members and channel it into one comprehensive viewpoint, articulated in one unified voice, it may actually be beneficial to have diverse input in the decision.*

6.4   SATISFICING

Another suboptimal decision-making process is satisficing. Herbert Simon named satisficing using an archaic English verb, satisfice, which meant to satisfy. It can best be thought of as having the combined meaning of the two modern verbs, satisfy and suffice. The goal of satisficing is not optimizing or maximizing a decision in the sense of achieving highest utility, probability, or logical soundness. The goal of maximizing is replaced during satisficing by that of finding a decision that satisfies all members of the group based on some set of minimal requirements—or even a single minimal requirement (Simon, 1956). The decision-making process stops when an option meets the minimum requirements of all associated parties. Satisficing commonly emerges in decision-making by heterophilic groups that are constrained from finding an optimum solution by the different objectives and values of its members. Satisfying the minimal needs of group members takes precedence over finding the optimum solution.[9]

> *One response to the concern with uncertainty, with the difficulties of discovering or designing alternatives, and with computational complexity has been to introduce search and information transmission processes explicitly into models. Another (not exclusive) response has been to replace optimization criteria with criteria of satisfactory performance. The satisficing approach has been most often employed in models where "heuristic" or trial-and-error methods are used to aid the search for plausible alternatives.*

Simon (1972)

The satisficing concept gradually became a serious alternative to decision-making aimed at optimization during the 1950s, rivaling groupthink in its eventual integration into the public psyche and vernacular (Fig. 6.2).

---

9. To avoid misrepresentation of this "heuristic," it must be pointed out that individuals such as executives also apply satisficing under other conditions. As an example, a professional investor might decide to purchase a stock if it meets some minimum requirement for profit (Janis & Mann, 1977). Their decision is not based on whether it produces the maximum profit.

There are advantages to the satisficing heuristic in some situations. Satisficing is the best alternative if heterophily precludes maximizing, or polythink precludes any decision. For individuals, satisficing also has the advantage of reducing postdecisional stress. Schwartz et al. (2002) found that individuals who tended to be satisficers when making personal decisions where generally happier and had higher self-esteem than individuals who maximized. Interestingly, this difference was not found for satisficing versus maximizing dyads making hiring decisions (Peng, 2013). Speculating, the sharing of responsibility of a decision by both dyad members might have reduced postdecisional anxiety and regret. Whatever the advantages of satisficing in certain situations, it is critical to keep in mind that maximizing the likelihood of picking the best option is not one of them.

To summarize our discussion to this point, groups with different degrees of homophily are prone to different errors (Fig. 6.3). Complete objectivity as described for science in Chapter 3, Human Reasoning: Within Scientific Traditions and Rules, should eventually result in a maximized choice based on existing evidence. There will be transient periods of polythink during any scientific endeavor attempting to find the best explanation for a set of evidence. If a group deviates from objectivity due to excessive focus on group coherence, groupthink emerges to produce a suboptimal choice. If the level of group heterophily is moderate or severe, deviations from a maximized choice could result from satisficing or polythink. Whether a maximized decision is necessary depends on the goal of the group. It is necessary for a group focused on a scientific issue but likely would be unobtainable by a risk assessment team working under stricter constraints.

6.5 ELICITING JUDGMENTS FROM GROUPS OF EXPERTS

The reader likely has already begun wondering whether elicitation of judgments from acknowledged experts might avoid many of the issues just described. The quick answer is yes in some instances, but no in others.

Reasoned, albeit still subjective, opinions of acknowledged experts on an issue are brought together during expert elicitation to generate an aggregate decision or estimate. Experts are identified based on their knowledge of the subject in hand, and importantly, also their skills in organizing and

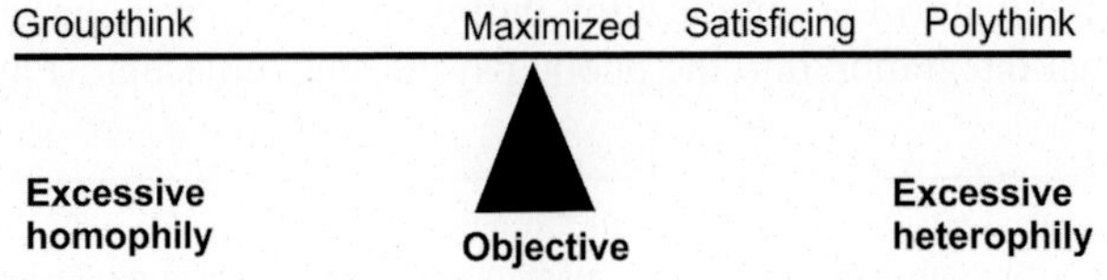

FIGURE 6.3 Summary of potential group decision errors in the context of homophily and heterophily.

applying evidence (O'Hagan et al., 2006). Different types of expert group-ings and tasks exist that strongly influence the process of expert elicitation. An expert panel for some elicitations might be configured as an SMT as described above. The associated experts would be no less susceptible to individual cognitive or group errors than any other collection of indivi-duals. Another elicitation type might involve an elicitor who independently asks each expert for their input and then composites the responses to pro-duce a final judgment. In this case, the exercise would simply be one of aggregating results from a series of decider–advisor dyads. During the pro-cess, involved individuals are subject to individual cognitive and dyad errors, but not group errors.

A more involved elicitation might intentionally mix elements of both dyad and group interactions as in the case of a Delphi expert elicitation (Helmer, 1968). Experts are chosen from among the most informed individuals in the topic area and each one given enough detail about the issue to begin their independent deliberations (Ayyub, 2001). Experts deliberate independently of one another in this initial step of a Delphi elicitation, responding directly to the elicitor who eventually combines their input with those from the other experts. The elicitor then sends the composite results back to all experts, allowing them to examine all responses and potentially modify their own judgments based on what they learned from the composite results. Additional rounds of reviewing composite results might be needed to arrive at a final consensus. The number of effective rounds is limited because Sackman (1975) found that convergence with too many rounds can result as much from boredom as true consensus. Dropouts before the elicitation's completion can also become problematic. Nonetheless, key improvements arising during the Delphi method are independence of experts during the first step and the abil-ity of experts to learn from the composite results and to refine their opinions.

[Plato's ideal state as described in] The Republic was governed by "guar-dians" who had undergone a lifetime of rigorous training in science, statecraft, and ethics ... [But] experts are not the guardians Plato had in mind. They are ordinary mortals with ordinary foibles.

Cooke (1991)

During a seminar by the above quoted expert-on-experts, Cooke made the unsettling statement that the correctness of an expert's opinion is not strongly correlated with their years of experience or their perceived standing in the field as gauged by the opinions of their peers (R. M. Cook, Personal Communication). Distressed by this gloomy conclusion and hoping that I had misunderstood, I asked him after his seminar if he had reached that con-clusion, and if so, under what conditions might it not be the case. After acknowledging that I had heard him correctly, he answered the second ques-tion with a pleasant smile and shoulder shrug. The resulting discomfort left a lingering question in my mind about the value of expert opinion elicitation.

Was the value only that of bolstering the outward appearance of legitimacy or credibility of a judgment?

Providing further insight about the nature of experts, Armstrong (1980) found that the accuracy of an individual is not a simple function of their level of expertise (Fig. 6.4). In terms introduced in Chapter 2, Human Reasoning: Everyday Heuristics and Foibles, individuals with minimal expertise must rely heavily on the recognition heuristic, resulting in considerable error or inaccuracy. As expertise increases, more and more "local" knowledge becomes available with which to make increasingly accurate predictions. Over a range of expertise levels beyond a certain minimum, enough "local" knowledge will be available to produce accurate predictions. Oddly enough, complacency of highly expert individuals can put them at a disadvantage. It can increase an expert's susceptibility to overconfidence and make them reluctant to considering alternatives or seeking out disconfirming evidence. The result is a decrease in accuracy despite increased levels of expertise. The message here is that accuracy rapidly increases and then plateaus as expertise increases. It might decrease beyond a certain point as highly expert individuals become increasingly susceptible to specific decision-making errors. Based on these observations, Armstrong (1980) formulated the seer-sucker theory that "no matter how much evidence exists

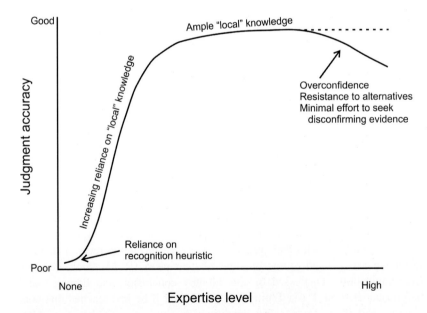

FIGURE 6.4 General trend in expert judgment or prediction accuracy as a function of level of expertise.
Modified from figure 1 of Armstrong, J.S. (1980). The seer-sucker theory: The value of experts in forecasting. Technology Review, 82(7), 16–24.

that seers do not exist, suckers will pay for the existence of seers." His point being that a moderately informed individual might be as or more accurate than an expensive, highly informed expert consultant. Often the motivation for engaging individuals with very high levels of expertise is simply enhancing credibility of the elicitation exercise, not increasing accuracy of estimates or predictions.

Expert elicitation might produce a qualitative or quantitative response. An example of a qualitative answer might be the recommendation to permit or prohibit the replacement of the ozone-depleting refrigerants, chlorofluourocarbons, with hydrochlorofluorocarbons that are less harmful to the ozone layer but act as potent greenhouse gases (Cornwall, 2016). A quantitative answer might be an estimate of the decrease in annual songbird fledgling production on a mercury-contaminated floodplain. Some quantitative answers can be probabilistic such as the one predicting the type of distribution and associated parameters for environmental mercury exposure (Wang & Newman, 2012). These different kinds of elicitations require slightly different approaches to be successful.

As indicated in Fig. 6.1 and Section 6.6.3, adequate expert independence, training, and calibration are crucial to a successful elicitation. Independence is important because the influence on an expert's answer due to interaction with the other experts on a panel creates the risk of the group errors described already. Realistically, individuals comprising a pool of experts will never be completely independent because they will share aspects of training and common knowledge from published literature on the subject of interest (Yaniv, 2004). Beyond independence, adequate training about the specific elicitation task makes certain that all experts have enough expertise so that they avoid reliance on the recognition heuristic. This can be especially critical when experts are being asked to provide distributional information as done in probabilistic studies because experts without formal probability calculus training can become victims of base rate neglect or probability blindness. Next, calibration of experts allows estimates from each expert to be weighted during final aggregation of all estimates.

Calibration can take several forms. Some elicitations present experts with questions for which most informed individuals in the field should know the answers. An expert's proportion of correct answers to calibration questions can be used as a weighting. For questions that require a quantitative answer, a similar approach might be taken but the closeness of an expert's answer to the true answer is used for weighting instead. The answers and weights for the experts are combined to generate a weighted aggregate estimate. In another approach, experts might be asked to indicate their answer and then provide an interval around that answer indicating their certainty. The intervals can then be used to weight answers during aggregation of expert estimates. As an example involving i expert estimates of an interval midpoint (x_i) and width (g_i), Yaniv (1997) gives a straightforward aggregated midpoint estimate from the individual midpoints and associated weights (w_i) of the experts,

$$\bar{x}_w = \sum_{i=1}^{n} w_i x_i.$$

Each weight (w_i) is generated using the expert's estimated interval width,

$$w_i = \frac{1/g_i}{\sum_{k=1}^{n} 1/g_k}.$$

These weightings might have been estimated after judicious removal of outliers. The challenge in doing this is avoiding the removal of any estimates based on sound, but unshared, insights or information. Yaniv (1997) provides good advice for trimming and weighting expert opinions.

6.6 REDUCING THE IMPACT OF GROUP ERRORS

6.6.1 Groupthink and Polythink

The chances of groupthink emerging during deliberations might be reduced by careful selection of group members before the decision-making process begins, that is, by thoughtful decision unit engineering.[10] As one example of a quality to consider, strongly committed individuals in a group are less prone to self-censor than are weakly committed members (Packer, 2010; Packer & Chasteen, 2010). Such an individual is committed more to the overall goal than to any particular option or group members. They are willing to pay the social cost of objecting if the decision-making process begins to go awry (Packer & Chasteen, 2010).[11] A key quality favoring inclusion of an individual is perceived strength of commitment to the overall objective. Past affiliation with other group members or published expertise with a particular potential decision option should also be carefully scrutinized because they can become the seed of groupthink if casually managed. As an example, my opinion is that the aforementioned EPA panel was victimized by groupthink because a subset of its members, including the leader, had successfully advocated together for the chosen method in past working groups. They quickly formed an in-group in this new panel that then drove

---

10. A decision unit is simply any group of individuals making the decision. See (Russett, Starr, and Kinsella, (2006) for formal details.

11. An individual's motivation to take on a group goal and internalize it will "engage in a process of depersonalization in which [individuals] take on the identity of and begin to act in accordance with the norms of the group..." (Packer & Chasteen, 2010). The individual begins to identify with the collective motives of the group. Groupthink can result under the wrong conditions but the right conditions will produce strongly committed individuals who readily object when the group's integrity is threatened.

the process toward their favored option. Suggesting a way of avoiding such conditions, Couzin et al. (2011) found that inclusion of moderately informed, but motivated, individuals in groups can curb the undue influence of other group members who have strong initial preferences. Couzin et al. modeled groups with two opposing subgroups of different relative sizes and intensities of bias, finding that the inclusion of unbiased individuals lessened the impact of biased subgroups on the eventual choice.

Janis (1971) outlined steps that can be taken to reduce the risk of groupthink after formation of the decision-making group (Table 6.3). The original list of recommendations was relevant to policy-making groups at the federal government level, but eight of the nine recommendations are also pertinent to ecotoxicology and ecological risk assessment groups that function as SMTs. Janis's seventh recommendation is not generally germane because it involves policy issues that require consideration of rival nation or organization policies.

Several themes are apparent in Janis's suggestions (Table 6.3). Following Suggestions 1 and 2 would make it clear that leaders must explicitly empower group members to be impartial and state that each member has an obligation to be critical when warranted. Several steps can avoid in-group insulation. One step is forming subgroups that deliberate and then come back together with their potential options. Another is inviting outside experts to periodically engage and challenge the group's thinking. Importantly, one or more formally designated devil's advocate should be assigned, and their roles and obligations to challenge the group clearly stated at the onset of deliberations. As an essential practice, the group should be given a break after an initial decision is formulated so that members can privately contemplate its basis and advantages relative to other options. Then the group should reconvene to present any additional insights or objections that might have emerged during this period of reflection.

Many readers of the above suggestions will conclude that "we do not have time for all this and need to just get on with making the important decision." Often, this is followed by, in my opinion, one of the most corrosive phrases commonly uttered in ecotoxicology and environmental risk assessment activities, that being that something is *"good enough."* When confronted with this disputable conclusion, I normally make two comments. The first is that I have seen many *"good enough"* decisions cause much confusion and harm later. These suboptimal decisions impede and confuse work done by future groups and individuals trying to reach sound decisions about related problems. The establishment of the now problematic null hypothesis significance test–based NOEC/LOEC metrics for chronic effects is used to illustrate this point. The second is to repeat the trite, yet true, adage "If you do not have time to do it right the first time when will you have time to do it over?" Using the NOEC/LOEC example again, I ask when will our field have the political will and time to discard all of the flawed NOEC/LOEC

TABLE 6.3 Remedies for Groupthink Outlined in Janis (1971) and Janis and Mann (1977)[a]

1. Before the proceedings begin, each person on the panel should be told by the group leader that their role is that of a critical evaluator who has the right to voice doubts or objections.
2. Those individuals in the organization who assign the task to the newly formed group should not state any preferences, and instead, should state their intent to be impartial. Rose (2011) further suggests that the group leader refrain from expressing their preference and focus instead on cultivating a legitimate consensus.
3. To avoid an adverse impact of possible in-group insulation, the organization should set up several independent groups led by different leaders to work on the same issue.
4. The group leader should require each member to discuss the group's thinking and tentative options with others outside of the group and then bring the feedback to the group for open discussion. This is to be done before any final decision being made by the group.
5. Outside experts should be invited to participate in relevant group meetings and challenge the views of core members.
6. At least one group member should be formally assigned the role of devil's advocate who challenges the group's emerging positions. This individual's role should be formally identified and their activities protected against any attempts at mindguarding.
7. Two or more subgroups with different leaders should at important junctures in the process be formed and then come together to resolve differences in their approaches or thinking.
8. When the group reaches an initial consensus, a break should be taken during which each member can reflect independently about it. A "second-chance" meeting is then held in which each member presents any objection, new insights, or information. A final decision is reached only after this opportunity is made available to reconsider or refine the decision.

[a]*One policy-related remedy from Janis's list was omitted here because it seemed irrelevant to ecotoxicology and ecological risk assessment.*

data and related decisions accumulated to date and replace them with sounder metrics of chronic effect.

Relative to polythink, it should be understood that it will appear at times in healthy scientific endeavors and does not reflect any failure in scientific decision-making. Multiple explanations for in-hand evidence are essential to eventually identify the hypothesis with the most explanatory power. Any attempt to "fix" a scientific endeavor by employing one of the mechanisms described below will compromise scientific progress. However, if the context is nonscientific, several ways exist for lowering the likelihood of polythink (Table 6.4).

Reducing polythink in an activity such as a formal ecological risk assessment begins by bringing together a decision unit with a good chance of

TABLE 6.4 Symptoms of Polythink as Described in Mintz and Wayne (2016)[a]

1. Contrasted with groupthink, the many different vantages on the subject and possible remedies exist that increases the probability of conflict and disagreements.
2. Reduced willingness to communicate among members emerges and an increase in mixed messages makes it difficult to define the best available options.
3. Framing of concepts and arguments increases as members positively present opinions favoring their set positions or frameworks, and negatively discuss other concepts and arguments.
4. A decision-making process often emerges that applies a lowest common denominator criterion for acceptability of a solution or decision (see Section 6.4.)
5. Decision paralysis occurs when even a lowest-common-denominator criterion for acceptability fails.
6. Review of available options is limited by the group which increases the risk of the best option being ignored.
7. Limited reflection of previous options after the group makes its judgment.

[a]*One symptom (likelihood of media leaks) relevant to the original conception of polythink in political decision-making was omitted because it was deemed irrelevant to most ecotoxicology and risk assessment teams.*

producing a high-quality decision. A good start is a team of motivated problem solvers with diverse perspectives who are committed to the common goal and agenda (Mintz & Wayne, 2016).[12] Also, a clear statement of the agenda and acceptable solutions would be essential.

After a group is formed, a skilled leader could turn polythink-generating conditions to the group's advantage by exploring the diverse options represented in the heterophilic group. Mintz and Wayne (2016) also suggest that polythink can create the opportunity to break an overarching problem into several smaller and more tractable problems.

6.6.2   Satisficing

... organisms adapt well enough to "satisfice"; they do not, in general, "optimize."

Simon (1956)

The above discussion suggested that groups afflicted with groupthink will come to suboptimal decisions. However, what constitutes an optimal decision was never explicitly stated. An optimal or maximized decision might be one derived by correct logic or estimation such as decisions

---

12. Perhaps, reducing polythink was the intent of the EPA organizers who picked the members of the panel described above as victimized by groupthink. They may have had a solution in mind (i.e., the solution already introduced into other regulatory activities by the group leader and a few of his colleagues) that was not declared to all group members at the onset.

generated with utility analysis (see Section 2.2.1 in Chapter 2: Human Reasoning: Everyday Heuristics and Foibles) or probability calculus (see Section 2.3.2.11 in Chapter 2: Human Reasoning: Everyday Heuristics and Foibles). This context is appropriate for many relevant situations, especially scientific inquiry. Groupthink, polythink, and satisficing are compromised group heuristics to be avoided in scientific inquiries although polythink might be characteristic at one stage of scientific inquiry. Nonetheless, satisficing might not be suboptimal in some activities such as ecological risk assessment in which the goal is the best achievable decision under the group dynamics and constraints. Even then, the deviation from maximization due to satisficing might be lessened by engaging a third-party mediator (Zartman, 2006).

6.6.3 Improved Expert Elicitation Techniques

Reducing error in expert elicitation begins with choosing the appropriate type and number of experts. Depending on the nature of information being elicited, as few as 3–10 experts are often adequate (Armstrong, 1980; Hogarth, 1978; Yaniv, 2004). Estimates of a simple average usually require fewer experts than estimates of more complete distributional information. Regardless, the number should be large enough to include a healthy diversity of "opinion, credibility, and result reliability" (Ayyub, 2001).

Selecting experts sharing a common experience, such as two experts who recently participated together on a committee addressing a related subject, would risk a violation of independence and probably produce little additional insight during aggregation of opinions (Yaniv, 2004). One of the similar individuals would likely provide nearly the same amount of useful insight as both of them together. In some cases, experts with known differences in vantages might be preferable (O'Hagan et al., 2006).

> ... the goals for any expert selection method should be to identify the appropriate expertise, to achieve a representation of scientific and ideological perspectives, and to do so in a process that is fair, reproducible, and reasonably transparent to outside observers.
>
> Walker, Evans, and MacIntosh (2001)

The certain amount of training will improve the process. Minimally, the reasons behind the elicitation and intended use of the results should be provided to each expert (O'Hagan et al., 2006). General presentation of common heuristics and biases that can befuddle judgment might also help some experts make better judgments. Training that encourages the habit of searching for disconfirming evidence for each decision option can also be advantageous (Armstrong, 1980). Additional training might be required for a collection of experts possessing different degrees of numeracy and probability estimation training. Otherwise, untrained experts asked to estimate probabilities—perhaps even probability density distributions—might fall back on global introspection, producing biased results (Théophile et al., 2010). Training experts in specific steps to take in generating probability information can greatly improve results (O'Hagan et al., 2006).

Approaches to calibration or weighing of experts differ among elicitations. The above example of using expert-provided ranges of confidence to weigh answers is one way. In a probabilistic study of songbird dietary exposure to mercury, Wang and Newman (2012) embedded questions within the associated expert elicitation that permitted the knowledge level of each expert to be determined. Answers to these calibration questions then allowed weighing of each expert's opinion.

Accurate predictions are possible with careful selection, training, and calibration of experts. Walker et al. (2001) and Walker, Catalano, Hammitt, and Evans (2003) did this with seven experts to generate magnitude and uncertainty estimates for human personal exposure to benzene. Experts were selected by peer nomination and participated in a 2-day training workshop. They also received a notebook containing papers relevant to benzene exposure. They attended a workshop that provided details on the project and discussed the common heuristics of anchoring, availability, overconfidence, and base rate bias. They provided inputs independently during 1.5 day face-to-face interviews with elicitors. Insights about probabilistic methods were provided by the elicitors, including publications, data, and simulations as requested by each expert. Expert calibration (Walker et al., 2003) involved comparison of each expert's judgment to actual measurements in the region of study (NHEXAS Region V). The overall results indicated successful expert calibration with the elicitation providing better estimates than those from compilations of measured benzene concentrations in other US cities.

6.7 CONCLUSIONS

The issues associated with decision-making by dyads and small groups were outlined and included naïve realism, groupthink, satisficing, and polythink. Additional challenges for expert elicitation included the seer-sucker fallacy, and inadequate expert independence, calibration, and training. Means of reducing the impact of these factors were described. Importantly, the seriousness of a group behavior such as satisficing depended on the decision-making context. Satisficing might be unacceptable in making a scientific judgment but the only acceptable way forward in some ecological risk assessments. Polythink is a necessary condition at some stages of a scientific inquiry, but it can be paralyzing for a group under a mandate to produce a judgment within a certain time period or with available evidence.

REFERENCES

Armstrong, J. S. (1980). The seer-sucker theory: The value of experts in forecasting. *Technology Review, 82*(7), 16–24.

Ayyub, B. M. (2001). *Elicitation of expert opinions for uncertainty and risks.* Boca Raton: CRC Press.

Baekkeskov, E. (2016). Explaining science-led policy-making: Pandemic deaths, epistemic deliberation and ideational trajectories. *Policy Sciences, 4,* 395–419.

Baron, R. S. (2005). So right it's wrong: Groupthink and the ubiquitous nature of polarized group decision making. *Advances in Experimental Social Psychology, 37,* 219–253.

Baron-Cohen, S., Leslie, A. M., & Frith, U. (1985). Does the autistic child have a "theory of mind. *Cognition, 21,* 37–46.

Bonaccio, S., & Dalal, E. S. (2006). Advice taking and decision-making: An integrative literature review, and implications for the organizational sciences. *Organizational Behavior and Human Decision Processes, 101,* 127–151.

Cooke, R. M. (1991). *Experts in uncertainty. Opinion and subjective probability in science.* New York: Oxford University Press.

Cornwall, W. (2016). How a figure key to new HFC pact was born. *Science, 354,* 402.

Couzin, L. D., Ioannou, C. C., Demirel, G., Gross, T., Torney, C. J., Hartnett, A., ... Leonard, N. E. (2011). Uninformed individuals promote democratic consensus in animal groups. *Science, 334,* 1578–1580.

Descartes, R. (2003). *A discourse on method and meditations* (E.S. Haldane & G.R.T. Ross, Trans.). Mineola, NY: Dover Publications. (Original work published 1637).

Harvey, N., & Fischer, I. (1997). Taking advice: Accepting help, improving judgment, and sharing responsibility. *Organizational Behavior and Human Decision Processes, 70,* 117–133.

Helmer, O. (1968). Analysis of the future: The Delphi method, and the Delphi method—An illustration. In J. Bright (Ed.), *Technological forecasting for industry and government.* Englewood: Prentice Hall.

Hogarth, R. M. (1978). A note on aggregating opinions. *Organizational Behavior and Human Performance, 21,* 121–129.

Janis, I. L. (1971). Groupthink. *Psychology Today, 5,* 43–46 & 74–76.

Janis, I. L. (1982). *Groupthink.* Boston: Houghton Mifflin.

Janis, I. L. (1989). *Crucial decisions: Leadership in policymaking and crisis management.* New York: The Free Press.

Janis, I. L., & Mann, L. (1977). *Decision making. A psychological analysis of conflict, choice, and commitment.* New York: The Free Press.

Kämmer, J. E., Gaissmaier, W., & Czienskowski, U. (2013). The environment matters: comparing individuals and dyads in their adaptive use of decision strategies. *Judgment and Decision Making, 8,* 299–329.

Klein, D. B., & Stern, C. (2009). Groupthink in academia: Majoritarian departmental politics and the professional pyramid. *The Independent Review, 13,* 585–600.

Knoke, D., & Yang, S. (2008). *Social network analysis* (2nd ed.). Los Angeles: SAGE Publications.

Kooijman, S. A. (1987). A safety factor for LC50 values allowing for differences in sensitivity among species. *Water Research, 21,* 269–276.

Kozlowski, S. W., & Ilgen, D. R. (2006). Enhancing the effectiveness of work groups and teams. *Psychological Science in the Public Interest, 7,* 77–124.

Kynn, M. (2008). The 'heuristics and biases' bias in expert elicitation. *Journal of the Royal Statistical Society. Series A, 171,* 239–264.

Liberman, V., Minson, J. A., Bryan, C. J., & Ross, L. (2012). Naïve realism and capturing the "wisdom of dyads. *Journal of Experimental Social Psychology, 48*, 507–512.

March, J. G. (1994). *A primer on decision making.* New York: The Free Press.

McCauley, C. (1998). Group dynamics in Janis's theory of groupthink: Backward and forward. *Organizational Behavior and Human Decision Processes, 73*, 142–162.

McPherson, M., & Smith-Lovin, L. (1987). Homophily in voluntary organizations: Status distance and composition of face-to-face groups. *American Sociological Review, 52*, 370–379.

McPherson, M., Smith-Lovin, L., & Cook, J. M. (2001). Birds of a feather: Homophily in social networks. *Annual Review of Sociology, 27*, 415–444.

Miller, N., Garnier, S., Hartnett, A. T., & Couzin, I. D. (2013). Both information and social cohesion determine collective decisions in animal groups. *Proceedings of the National Academy of Sciences of the United States of America, 110*, 5263–5268.

Mintz, A., & Wayne, C. (2016). The polythink syndrome and elite group decision-making. *Advances in Political Psychology, 37*, 3–21.

Mlodinow, L. (2012). *Subliminal. How your unconscious mind rules your behavior.* New York: Pantheon Books.

Moorehead, G., Neck, C. P., & West, M. S. (1998). The tendency toward defective decision making within self-managing teams: The relevance of groupthink for the 21st century. *Organizational Behavior and Human Decision Processes., 73*, 327–351.

Neck, C. P., & Manz, C. C. (1994). From groupthink to teamthink: Toward the creation of constructive thought patterns in self-managing work teams. *Human Relations, 47*, 929–952.

O'Hagan, A., Buck, C. E., Daneshkhah, A., Eiser, J. R., Garthwaite, P. H., Jenkinson, D. J., … Rakow, T. (2006). *Uncertain judgements. Eliciting expert's probabilities.* Chichester: John Wiley & Sons, Ltd.

Packer, D. J. (2010). Avoiding groupthink. Whereas weakly identified members remain silent, strongly identified members dissent about collective problems. *Psychological Science, 20*, 546–548.

Packer, D. J., & Chasteen, A. L. (2010). Loyal deviance: Testing the normative conflict model of dissent in social groups. *Personality and Social Psychology Bulletin, 36*, 5–18.

Peng, S. (2013). Maximizing and satisficing in decision-making dyads. *Wharton Research Scholars Journal.* Paper 98. http://respository.upenn.edu/wharton_research_scholars.

Premack, D., & Woodruff, G. (1978). Does the chimpanzee have a theory of mind? *The Behavioral and Brain Sciences, 4*, 515–526.

Price, P. C., & Stone, E. R. (2004). Intuitive evaluation of likelihood judgment producers: Evidence of a confidence heuristic. *Journal of Behavioral Decision Making, 17*, 39–57.

Rauwolf, P., Mitchell, D., & Bryson, J. J. (2015). Value homophily benefits cooperation but motivates employing incorrect social information. *Journal of Theoretical Biology, 367*, 246–261.

Richerson, P. J., & Boyd, R. (2005). *Not by genes alone. How culture transforms human evolution.* Chicago: University of Chicago Press.

Rose, J. D. (2011). Diverse perspectives on the groupthink theory—A literary review. *Emerging Leadership Journeys, 4*, 37–57.

Russett, B., Starr, H., & Kinsella, D. (2006). *World politics: The menu for choice.* Boston: Wadsworth.

Sackman, H. (1975). *Delphi critique, expert opinion, forecasting and group processes.* Lexington: Lexington Books.

Schwartz, B., Ward, A., Lyubomirsky, S., Monterosso, J., White, K., & Lehman, D. R. (2002). Maximizing versus satisficing: Happiness is a matter of choice. *Journal of Personality and Social Psychology.*, *83*, 1178–1197.

Simon, H. A. (1956). Rational choice and the structure of the environment. *Psychological Review*, *63*, 129–138.

Simon, H. A. (1972). In C. B. McGuire, & R. Radner (Eds.), *Theories of bounded rationality.* Amsterdam: North-Holland Publishing Company.

Sprong, M., Schothorst, P., Vos, E., Hox, J., & Van Engeland, H. (2007). Theory of mind in schizophrenia. *British Journal of Psychiatry*, *191*, 5–13.

Strom, D. J., Stansbry, P. S., & Porter, S. W. (1995). *Conflicting paradigms in radiation protection: 20 questions with answers from the regulator, health physicist, the scientist, and the lawyers. PNL-SA-24763.* Richland, WA: Pacific Northwest Laboratory.

Théophile, H., Arimone, Y., Miremont-Salamé, G., Moore, N., Fourrier-Réglat, A., Haramburu, F., & Bégaud, B. (2010). Comparison of three methods (consensual expert judgement, algorithmic and probabilistic approaches) of causality assessment of adverse drug reactions. *Drug Safety*, *33*, 1045–1054.

Turner, M. E., & Pratkanis, A. R. (1998). Twenty-five years of groupthink theory and research: Lessons from the evaluation of a theory. *Organizational Behavior and Human Decision Processes*, *73*, 105–115.

Tversky, A., & Kahneman, D. (1974). Judgment under uncertainty: Heuristics and biases. *Science*, *185*, 1124–1131.

Walker, K. D., Catalano, P., Hammitt, J. K., & Evans, J. S. (2003). Use of expert judgment in exposure assessment. Part 1. Calibration of expert judgments about personal exposures to benzene. *Journal of Exposure Analysis and Environmental Epidemiology*, *13*, 1–16.

Walker, K. D., Evans, J. S., & MacIntosh, D. (2001). Use of expert judgment in exposure assessment. Part 1. Characterization of personal exposure to benzene. *Journal of Exposure Analysis and Environmental Epidemiology*, *11*, 308–322.

Wang, J., & Newman, M. C. (2012). Projected Hg dietary exposure of 3 bird species nesting on a contaminated floodplain (South River, Virginia, USA). *Integrated Environmental Assessment and Management*, *9*, 285–293.

Yamamoto, H., & Matsumura, N. (2009). *Optimal heterophily for word-of-mouth diffusion. Proceedings of the Third International ICWSM Conference.* Palo Alto: Association for the Advancement of Artificial Intelligence.

Yaniv, I. (1997). Weighting and trimming: Heuristics for aggregating judgments under uncertainty. *Organizational Behavior and Human Decision Processes*, *69*, 237–249.

Yaniv, I. (2004). The benefit of additional opinions. *Current Directions in Psychological Science*, *13*, 75–78.

Zartman, I. W. (2006). Negotiating internal, ethnic and identity conflicts in a globalized world. *International Negotiation*, *11*, 253–272.

Chapter 7

How Innovations Enter
and Move Within Groups

7.1 INNOVATION DIFFUSION

During the first two decades of my professional career, I engaged diligently
at each annual meeting of the Society of Environmental Toxicology and
Chemistry, often serving as a formal judge of presentations. New scientific
approaches and ideas would intermittently appear at these meetings and
then grow increasingly popular after an initial lag period. Oddly, other
equally meritorious innovations would appear but invoke only mild interest
from scientists and regulators before fading into obscurity within a few
years.

Differences in the fates of innovations were also apparent in the peer-
reviewed literature. Some novel approaches or concepts drew considerable,
sustained interest from the scientific or regulatory communities, whereas
others did not. Through the years, I also noticed an occasional sleeping
beauty (Ke, Ferrara, Radicchi, & Flammini, 2014; Van Raan, 2004), that is,
a publication reporting novel insights or approaches that went virtually unnoticed for long periods before suddenly gaining much attention. The first
sleeping beauty in ecotoxicology that I noticed was global distillation and
fractionation (Goldberg, 1975a).[1] This concept lay dormant for approximately
20 years before a large number of scientists realized its relevance and published on the topic. Perhaps not coincidentally, the increased interest
occurred simultaneously with the surge in regulatory agencies' evidence
gathering before the 2001 Stockholm Convention for the Control of
Persistent Organic Pollutants. At the opposite end of the spectrum from

---

1. Global distillation is the process by which persistent organic pollutants (POPs) such as
volatile organochlorine compounds are distilled from warmer regions of use to cooler regions of
the globe. The phenomenon, global fractionation co-occurs because POPs differ in their individual rates of degradation, vapor pressures, and lipophilities. A fractionation occurs when some
POPs move more readily than others toward the polar regions. The net result is a redistribution
of the different POPs from the Equator or site of origin toward the cold polar regions of the
Earth (see also Appendix 1: Ecotoxicology Innovation Survey Methods).

The Nature and Use of Ecotoxicological Evidence. DOI: https://doi.org/10.1016/B978-0-12-809642-0.00007-8
185

sleeping beauties, some once-useful concepts and techniques seemed to persist well beyond their serviceable period. The still-pervasive applications of LC50 and the NOEC metrics for predicting short- and long-term exposure consequences to ecological populations and communities are classic examples of once-useful approaches that are now impediments to making sound ecological predictions. Such applications are intellectually indefensible pseudo-ecotoxicology, yet ecotoxicologists and ecological risk assessors alike persist in using them for this purpose.

Plausible reasons for differences in innovation introduction and persistence in our field began to absorb my thoughts as much as the scientific content of presentations and publications. This chapter sketches out some partial answers to the nagging question, what other than scientific soundness influences the fate of an innovation introduced into ecotoxicology or ecological risk assessment? For the first time in this book, this chapter and Chapter 8, Evidence in Social Networks, address issues from the vantage of emergent dynamics and structures in the science of ecotoxicology and the technology of ecological risk assessment.

> ... scientific advance depends not only on the rate of discovery but also upon diffusion, evaluation and use of the discoveries. Given the stratified and clustered nature of scientific communities, the nature of personal influence and information links becomes crucial to scientific exchange.
>
> Duncan (1974)

Two groups of academicians, science historians/philosophers and sociologists, examined how new ideas enter into a science like ecotoxicology or a group such as an ecological risk assessment team (Mulkay, 1974). Science historians described how paradigms seemed to come into being, compete with other paradigms, perhaps dominate for a period, and inevitably be displaced by another paradigm with superior explanatory power. Because of their interest in important historical changes, they focused more on major paradigm shifts than subordinate concepts or minor innovations. Their findings are most relevant to the science of ecotoxicology and less so to ecological risk assessment.

Beginning in the first half of the 20th century, sociologists more broadly explored the diffusion of novel concepts, beliefs, or technologies into groups, including groups of scientists or regulators. Early studies tried to determine what influenced the acceptance or rejection of a beneficial innovation by a group. For instance, the first innovation diffusion study (Ryan & Gross, 1943) focused on adoption by Iowa farmers of a novel hybrid seed corn. After a decade of mild interest in innovation diffusion, the approach used in this pioneering study inspired a flurry of similar studies by rural sociologists, firmly establishing the current innovation diffusion paradigm (Valente &

Rogers, 1995).[2] Approximately 4000 innovation diffusion studies had been published by 1995 (Valente & Rogers, 1995); a quick (May 23, 2017) search on the Web of Science database returned 9654 publications on the topic, innovation diffusion. Modern applications of innovation diffusion theory embrace a wide range of goals including enhancing market penetration of commercial products, understanding topic movement through social media, improving hygiene in rural villages, and discouraging illegal drug use in at-risk groups. This broad sociological vantage offers insight about innovation in activities such as pollution control, ecological risk assessment and scientific research. As pertinent examples, Popp (2010) applied diffusion theory to understand power plant adoption of nitrogen oxides (NOx) control technologies, and Loh (2011) used them to explain the transfer of environmental protection experiences from Hong Kong into mainland China.

Elements from both the science history and sociology vantages are blended into this chapter. First, the framework developed by science historian, Thomas Kuhn (1972, 1977, 2000) and elaborated upon by many others (e.g., Campanario, 1993; Crane, 1972; Fuchs, 1993; Majchrzak, Cooper, & Neece, 2004; Rzhetsky, Foster, Foster, & Evans, 2015) is presented and illustrated with ecotoxicological innovations. (See also our previous discussions of Kuhn's theory of scientific revolutions in Chapters 2 and 3.) Next the sociologists' vantage is outlined, especially that brought into focus by Rogers (1995) and enriched by many others (e.g., Granovetter, 1978; Mahajan & Peterson, 1985; Strang & Meyer, 1993; Valente, 1996; Wejnert, 2002). A brief discussion then follows about qualities of information transmission, those being specifically, integrity and fruitfulness (Morgan, 2011). Finally, a few ecotoxicological innovations will be analyzed from these vantages.

Regardless of whether the historical or sociological vantage is taken, the successful movement of most novel ideas or technologies into a group involves two crucial steps (Klein & Knight, 2005; Majchrzak, et al., 2004) (Fig. 7.1). Majchrzak et al. (2004) use the terms, acquire and integrate, and Klein and Knight (2005) use the terms, adoption and implementation for these two steps. Both sets of terms seem appropriate and are used herein. The innovation must first be acquired from another group. This might involve a group member searching the literature of another group or perhaps becoming actively engaged with members of a potential donor group. As an example, an individual might discover an approach with potential value to

---

2. Such a community or network of researchers associated with a common research paradigm or theme is referred to as an invisible college. Together, its members advance the science surrounding that central paradigm. According to Crane (1972),"Under the leadership of one or two scientists, the groups of collaborators [making up the invisible college] recruit and socialize new members ... the leaders of these groups also define the important problems for research in their fields ... their interpretations of seminal works influence the subsequent research of others in the field."

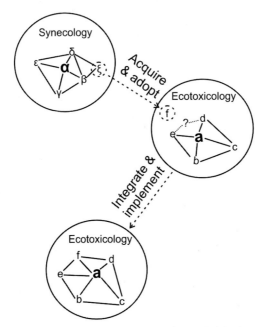

FIGURE 7.1 Two essential stages of innovation. An action or decision is made during the first stage to attempt adoption of the innovation. Klein and Knight (2005) refer to this as the adoption stage and Majchrzak et al. (2004) apply the label, acquisition stage. In the example shown here, a satellite concept or fact, ξ, associated with a synecological core paradigm, α, is identified as potentially useful in supporting the ecotoxicology core paradigm, a. Perhaps an ecotoxicologist exposed through her university department interactions to the synecology paradigm α, realizes the potential utility of ξ for interpreting correlations between d and e noted in ecotoxicology studies of a. The ξ is explored as a tentative explanation and perhaps customized to a slightly different formulation, f, during the implementation (Klein & Knight, 2005) or integration (Majchrzak et al., 2004) stage. The innovation becomes more firmly embedded in ecotoxicology. In this example, the innovation (f) did eventually support the core paradigm, a, and provided explanation for the heretofore inexplicable correlation between ancillary facts/concepts d and e.

ecotoxicology while attending a North American Benthological Society or a Society of Toxicology meeting. An ecological risk assessment professional might discover a novel way of identifying probable causes of an observed impact while attending a Society of Epidemiologic Research meeting.

As already mentioned in Chapter 3, Human Reasoning: Within Scientific Traditions and Rules, the innovator who introduces a radically new paradigm or approach does so at some risk to their status within their ingroup if the innovation threatens anyone vested in an establishing paradigm or technology. As a consequence, innovators are sometimes subject to active attempts by others to reduce their status and influence within ingroups (Campanario, 1993; Kirton, 1976; Kuhn, 1970; Nissani, 1995). A pertinent example is the initial difficulty that R. A. Fisher faced when trying to publish his now-classic 1918 statistics paper (Fisher, 1918). Also, due to the iconoclastic nature of his publications, he confronted difficulties finding employment early in his career and was reluctantly offered a position in Karl Pearson's well-established group "on the condition that he only publish what Pearson approved" (Nissani, 1995). (Karl Pearson had been one of the two reviewers who did not recommend publication of Fisher's classic 1918 paper). Incongruously, Fisher treated others in a similar fashion once he became an opinion leader. This was obvious in his response to Berkson's proposal to apply log logistic models to dose—response data. The log normal (probit) model for dose—response data had previously been introduced by Bliss (1935) and refined by Finney (1947) and Gaddum (1953) with significant input from Fisher.[3] Berkson (1951, 1953) proposed later that the (now widely accepted) log logistic model can be superior to the log normal model. This prompted a January 30, 1954, letter from Fisher to Berkson containing the following dismissive sentence, "[Your statements] would seem to involve comparisons of three different kinds and so may mean nothing logically, and have been made purely for its propagandist effect." Clearly, Fisher had learned the wrong lesson from the prior suppression of his own ideas by others.

The second step toward successful diffusion of an innovation is its implementation or integration with existing paradigms, technologies, and activities of the group. Implementation is "the transition period during which individuals ideally become increasingly skilled, consistent, and committed in their use of the innovation" (Klein & Knight, 2005). Without effective implementation, Ralph Waldo Emerson's adage is demonstrably false that "If a man can write a better book, preach a better sermon, or make a better mousetrap than his neighbor, though he build his house in the woods, the world will make a beaten path to his door."

The crucial integration or implementation step can fail for one or more of the following reasons (Klein & Knight, 2005). The last four reasons would seem especially relevant to environmental techniques such as those mandated by regulatory agencies in ecological risk assessment.

- Many newly acquired innovations are imperfectly suited to their new purpose and require some adjustment. Their appropriateness or utility is often questioned during this adjustment period.
- Many innovations require users to develop new skills and accept novel logical constructs, causing some stress and dissonance during implementation.
- In formal institutions such as many regulatory agencies, the decision to implement an innovation is made by individuals high in the hierarchy but implementation is carried out by those below these decision makers.

3. The engagement of Fisher in this invisible college is evident from the following comment from the Foreword of Finney (1947), "Much of this change [in the acceptance of statistical analysis of toxicity data] is due to the school of statisticians founded by R. A. Fisher ...."

Users can require some persuasion before abandoning an existing concept or technology, and implementing the innovation.

- Individuals in an organization often are required to change their role or function for the institution to accommodate the innovation. This requires careful thought and skillful coordination to be successful.
- Converting from one concept or technology to a new one requires expenses including time, retraining, money, and increased management activities. Inadequate expenditure of resources and time can increase the risk of implementation failure.
- By their nature, organizations serve as stabilizing forces. For example, the EPA might provide stability by specifying acceptable standard methods and approaches. The replacement of an existing concept or technology with a new one requires a transitory lessening of the stabilizing influence of an organization.

7.2 KUHN'S THEORY OF SCIENTIFIC REVOLUTION

话说天下大势，分久必合，合久必分.

(The empire, long divided, must unite; long united, must divide. Thus it has ever been.)

Attributed to Luo Guanzhong (1321, trans.1995)

Thomas Kuhn's (1970) book, *The Structure of Scientific Revolutions*, scrutinized major historical scientific discoveries and proposed a general pattern or life cycle for discoveries. First, facts and evidence are collected on a new topic. Next, an explanation for the observations is proposed. Further evidence gathering occurs using this new explanatory principle or paradigm, as the core around which acceptable kinds of studies and surveys are designed.[4] As evidence accumulates and is evaluated from the context of the paradigm, some will eventually be uncovered that are incongruous. As inconsistent evidence accumulates, the opportunity emerges for a new paradigm with higher explanatory power to displace the existing one. The successful new paradigm will become the core around which future studies are designed and evidence explained. The process continues through a notionally unending progression of increasingly more powerful paradigms. Kuhnian paradigms emerge, persist as favored explanations for a period, and then slowly disappear as more satisfactory paradigms displace them.

In my opinion, two clear examples of failing paradigms exist in ecotoxicology today. The first is that adverse ecological impacts of contaminants

---

4. Paradigms in a formal Kuhnian sense are "... universally recognized scientific achievements that for a time provide model problems and solutions to a community of practitioners" or "... law, theory, application, and instrumentation together—[that] provide models from which spring particular coherent traditions of scientific research" (Kuhn, 1970).

can be adequately predicted from collections of individual-based effects metrics. Chronic criticism of this paradigm exists in the literature, yet it continues to be a central theme around which many studies are designed and executed. The second is that strong, objective, and defensible inferences can be made with current null hypothesis significance test–based data evaluation procedures. Openly debated in the context of the NOEC metric, broader questioning of this paradigm has only just begun in the ecotoxicology literature. As suggested in Chapter 5, Individual Scientist: Reasoning by the Numbers, this paradigm is coming under increased questioning and more powerful paradigms are emerging to displace it.

As an example of a possible emerging ecotoxicology paradigm, MacDonald, Mackay, and Hickie (2002) introduced the novel conceptualization called environmental amplification that encompasses bioconcentration, bioaccumulation, and biomagnification. Their framing of these three processes around one paradigm is based on the two sequential steps of solvent switching and solvent depletion. Relative to POP biomagnification, amplification involves partitioning of the contaminant from water to the lipids in prey tissues (solvent switching). This is followed by predator digestion of prey lipids (solvent depletion) with partitioning of the contaminant into predator tissue lipids. Solvent switching and depletion can also be used to frame POP biocentration and bioaccumulation. It remains unclear at this time whether this new paradigm will supplant the distinct and more descriptive paradigms for bioconcentration, bioaccumulation, and biomagnification. If it does, another desirable goal in science would be served, that is, theory reduction in which several concepts are integrated into one.

Perhaps, because the science of ecotoxicology is so young, my explorations of the associated literature uncovered no ecotoxicology innovations that had enough time to appear, peak, and then wane in accordance to Kuhn's model. The best I could do was to use the pre-ecotoxicology example of the saprobic index. This index was based on the saprobien system for characterizing degraded aquatic communities that develop at various distances below discharges of putrescible organic waste (Kolkwitz & Marsson, 1908; 1909). The saprobien index was pertinent to the then-plentiful, slow flowing rivers receiving untreated sewage at the start of the 20th century (Chutter, 1972). It was formalized by Pantle and Buck (1955) and later linked to waterbody 5-day biological/biochemical oxygen demand by Sládeček and Tuček (1975). Introduced in the mid-1950s, the saprobien index was modestly cited in the published literature until c. 1990 (Fig. 7.2). Its appearance in the literature then increased until approximately 2010, at which time citations leveled off and began declining. It is still cited in the literature because recent applications have combined it with other metrics to create tools for gauging the general condition of flowing waters. One such tool is the stream assessment system built to fulfill the needs of the 2000 EU Water Framework Directive (Hering, Moog, Sandin, & Verdonschot, 2004). The emergence of

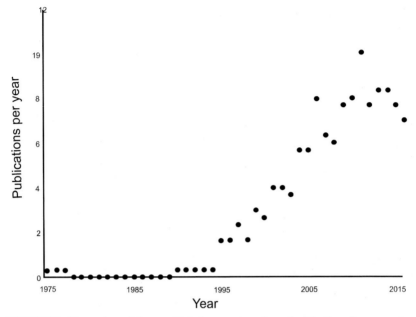

FIGURE 7.2 Time course of the saprobic index annual numbers of publications (3-year running averages shown) since 1975. The life cycle of a Kuhnian paradigm is suggested from the initially slow, accelerating, plateauing and declining of the annual numbers of publications on the topic. This index, designed to reflect the impact of putrescible organic pollutants, is gradually being subsumed by several multimetric biotic indices that also reflect effects of other kinds of contamination. *Produced with the Web of Science database tool, searched November 21, 2016.*

multimetric biotic indicators and the diminished emphasis on sewage-associated putrescible organic pollutants slowly shifted interest away from the saprobic index itself toward more complex indices.

Many of the details supplied in Kuhn's classic *The Structures of Scientific Revolutions* were framed around inexorable intellectual conflict including the active stymying of an innovator's efforts by those vested in an established concept or technique. Kirton (1976) describes the situation in slightly less combative terms, that is, there are always those seeking to "do things better" with existing paradigms or techniques (adaptors) and those seeking instead to "do things differently" (innovators). In the spirit of Kirton (1976), I would argue that these different modes of doing science do not always involve conflict as suggested by Kuhn. The rise and current decline in the saprobic index did not seem to have involved outright conflict. The two failing ecotoxicology paradigms just described are among the few threatened major paradigms associated with outright conflict. Large groups of ecotoxicologists and risk assessors are strongly committed to studies based on these paradigms. Despite these important exceptions, unbiased retrospection of transitions in ecotoxicology suggests that most innovations were not especially disruptive and conflict-inducing.

Some innovations do not elicit conflict because they present minimum threat to anyone's career or current endeavors. Unlike the major innovations studied by Kuhn, many are not radical enough to warrant protest or are gradual enough to be accommodated by most. Majchrzak et al. (2004) make the following distinction between radical and incremental innovation, "Radical innovation is differentiated from incremental innovation by involving discontinuous development where unprecedented improvements or performance features are achieved." Only mild resistance might be expected for an incremental innovation that moves the field ahead without seriously threatening any members of the scientific community. Most innovators risk less than suggested earlier in this book if their innovations are incremental. The conceptual and technical evolution discussed in Section 7.5.2.1 regarding the gill surface interaction model (GSIM), biotic ligand model (BLM), free ion activity model (FIAM), and diffusive gradients in thin films (DGT) complex is a good example.

In theory, radical innovations might also be integrated into our discipline by thoughtful and right-minded members without inciting excessive conflict. As most scientists are intellectually fascinated with and committed to their field of study, a new way of looking at or doing things might more often be welcomed as an opportunity for intellectual growth than resisted petty-mindedly as an impediment in one's career. For example, I vividly remember the first time I heard a presentation on global distillation and fractionation. It literally took my breath away for a moment when I suddenly grasped the concept and its implications. Never once I did think that it might draw funding and focus away from my professional research theme or those of friends. I suggest that most people in the sciences experience such moments but these experiences are not discussed as freely as the scientific gossip surrounding disagreements and conflicts. As a less personal example, the above story of Fisher's insinuendo about Berkson's logistic model can easily be reframed into one emphasizing Berkson's praiseworthy, nonconfrontational response that eventually resulted in the acceptance of both, the log logistic and log normal models for modeling dose−response data. Some poor behavior occurred but the community of scientists at large acted thoughtfully and with good results.

Rzhetsky et al. (2015) observed that a choice heuristic is often applied in scientific disciplines, that is, any particular scientist's willingness to accept a new concept or technique is strongly influenced by the beliefs of others with whom they interact. A quick review of Chapter 2, Human Reasoning: Everyday Heuristics and Foibles, will reveal that this "choice heuristic" is nothing more than a predictable manifestation of the imitation heuristic. The imitation heuristic operating within a social network produces differences in innovation acceptance within subgroups making up the larger scientific community.[5] Differences dictate the speed with which a belief moves through the entire social network: they are not reflective of conflict per se between innovators and those working with the current paradigms. Some resistance is indicative of healthy scrutiny and natural innovation diffusion dynamics, not petty conflict initiated by vested individuals.

Exploring various features of Kuhn's theory, Fuchs (1993) also concludes that there are other contexts in which science is conducted than the archetypical normal versus innovation science context. Some will involve little tension between those vested in old paradigms and those advocating new ones. The emergence of more collaborative science has added another feature to consider in understanding how innovations diffuse in a discipline. Uzzi, Mukherjee, Stringer, and Jones (2013) recently analyzed 17.9 million scientific publications, finding a common pattern in which teams with diverse backgrounds brought established concepts from their respective disciplines together to produce a scientific breakthrough. Clearly, team acceptance of the pieces transplanted from members' separate disciplines accelerates acceptance of innovation and reduces resistance.

7.3   INNOVATION DIFFUSION THEORY

A new scientific truth does not triumph by convincing its opponents and making them see the light, but rather its opponents eventually die, and a new generation grows up that is familiar with it.

Max Planck (1949)

Some adages seem designed more to comfort than inform. In my experience, Planck's above opinion is invoked most often to console someone whose cherished "truth" withered under the scrutiny of the scientific community. It deflects attention away from dispassionately determining whether the proposed "truth" was sound. I recollect reciting it myself on several occasions when a fleetingly favored opinion of my own was ignored in committee or the published literature. It minimized ego bruising for the moment needed to recollect that everyone carries the obligation to consider all evidence dispassionately, particularly that endangering a favored opinion. After an instant of reflection, I usually discard Planck's adage in favor of one adopted from Chamberlin (1897), "Own the process of doing excellent science, not a particular idea or position."

---

5. As discussed in more detail in Chapter 8, Evidence in Social Networks, the network structure of a scientific discipline relative to acceptance or implementation of new evidence, concepts, or paradigms typically takes on a core-and-periphery configuration (Duncan, 1974). One or a few clusters of scientists engage in most of the research on a theme with a loose scattering of other scientists in the community researching the theme in a less coordinated manner. Naturally, there will be different rates and degrees of acceptance of a novelty within such a structure. These differences are not always a result of petty conflict among those vested to different degrees.

Kuhn argued that Planck's observation was consistent with his theory of scientific revolution but admitted also that the cliché needed closer study (Kuhn, 1970). Further scrutiny suggested that important evidence is overlooked by those parroting the adage. Zuckerman (1996) examined Nobel laureates (1910−72), finding the mean age at which a scientist conducted their award winning research was 39 years old. Wray (2003) examined the 28 instances of scientific revolution cited in Kuhn's *The Structure of Scientific Revolutions* and found that the median age of scientists when they made their discoveries was 37 years old. These typically late early to middle career ages do not support the claim that innovation is the purview of the young scientist and that old scientists are obstructions to progress. Perhaps many innovations proposed by young scientists and resisted by those seasoned in the field are simply bad ideas. Or perhaps middle-aged scientists have enough insight and professional standing to risk formulating and then publishing an innovation. To me, the dubious soundness of Planck's adage suggests that the insights provided by Kuhn's "revolution model" do not fully encompass all innovative science. Many alternative insights pertinent to scientific innovators can be drawn from the sociology of innovation diffusion literature.

7.3.1   Basic Concepts and Qualities Influencing Diffusion

As used here in the context of scientific innovation, innovation diffusion is "the spread of abstract ideas and concepts, technical information, and actual practices within a social system, where the spread denotes flow or movement from a source to an adopter, typically via communication and [social] influence" (Wejnert, 2012). The rate at which an innovation is adopted depends on several features (Rogers, 1995): the qualities of innovation itself, type of required innovation-decision, available communication channels, nature of the social system, and the extent to which the change agent promotes the innovation.

Key qualities of the innovation include (1) its perceived advantage relative to the *status quo*, (2) its compatibility (perceived consistency with existing knowledge/paradigms, evidence, and needs of potential adopters), (3) perceived complexity or difficulty in implementation, (4) trialability, that is, the perceived ease with which it can be used on a trial basis before a decision to adopt must be made, and (5) observability, or how easily a potential adopter can see its advantages (Rogers, 1995). Adoption is accelerated if the innovation is obviously superior to that which is currently in place, compatible with existing theory and techniques, easily implemented, capable to being tested without much investment before full adoption, and has virtues obvious to potential adopters. Perception is often as critical as the physical reality of the situation.

Relative to types of innovation-decisions, the decision to adopt the innovation might be an optional one made by a single individual, in which case the adoption process can be relatively quick. Occurring at a slower rate are adoptions associated with a collective like a government agency. According to Rogers (1995), an adoption by a collective can be hastened by creating an authority framework in which only a small subgroup decides and the rest of the group must then comply with their decision.

The way in which information diffuses into the pool of potential adopters is also important. Mass media might be important for some innovations, especially straightforward innovations with no difficult nuances. Publication in a widely read, high-impact journal falls into the category of "mass media" for a science. Additional person-to-person transmission might be needed for innovations that are complex (Rogers, 1995). Individuals skilled relative to the innovation might visit a potential adopter's laboratory to train personnel about the details of successful application. Workshops might be held to train individuals on the mathematics, computational modeling, or statistics associated with a complex approach.

Presentations at annual meetings result in quicker diffusion initially than publication in the peer-reviewed literature because the peer-review and publication steps might take a year or more. An initial delay can be especially problematic for innovative science because the time to successfully pass the anonymous review process is often longer for an innovative study than for a normal science study (Campanario, 1993). As a counterpoint, the risk from too aggressively promoting an innovation before its careful review by peers can lead to an inferior innovation getting a toehold that might require much effort later to dislodge.

As discussed on Chapter 8, Evidence in Social Networks, features such as the interconnectedness of individuals within the social network or the number of connections an innovator has in the network are also important. Chapter 8, Evidence in Social Networks, will add more details about network structure and innovation diffusion.

The crucial factor identified by Rogers is the effort that an individual change agent expends to promote the innovation. Critical early in the adoption process is the expenditure of considerable effort by the change agent to gain the support of key opinion leaders. Opinion leaders are those in the group most likely to be imitated; however, recollect that they also tend to be general conformers to group norms. They might be resistant to change, and extra effort would be required to win them over. Aggressive promotion is less important once the opinion leaders endorse the innovation.

The dynamics of innovation diffusion generate curves that approach some maximum number of adopters with time (Fig. 7.3). The shape, including perhaps an initial long lag, is the net result of exposure of potential adopters to the innovation and subsequent decisions by individuals after exposure. Each individual weighs the advantage of the innovation, the amount of effort and resources that the adoption might require, and the degree of uncertainty associated with the innovation if adopted (Mahajan & Peterson, 1985). The curve shape results from the manner in which the innovation is introduced to individuals and the distribution of interindividual differences in proneness to adopt.

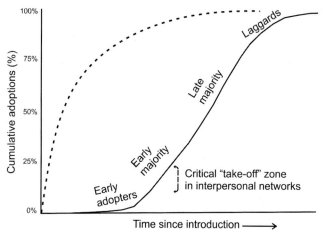

FIGURE 7.3   General curves of the cumulative percentage of adopters through time. For the internal influence model,"... diffusion rates first rise and then fall over time, leading to a period of relatively rapid adoption sandwiched between an early period of slow take up and late period of slow approach to satiation" (Geroski, 2000) (solid line). In some cases, there might be an additional lag or period of delay (Granovetter, 1978; Valente, 1996). Rogers (1995) indicates that adoption by 10%–25% of potential adopters is a critical mass needed for the innovation to quickly spread through an interpersonal network. Unlike the internal influence model, the external influence model (dashed line) produces a smooth curve that gradually converges on an asymptotic value.

A threshold might exist if there is a distinct subset of individuals who adopt much more readily than others who require a substantially large number of others to adopt before they are willing to adopt. The result is a time lag prior to reaching that critical number of individuals needed by the larger subgroup of potential adopters to make their decision (Valente, 1996). Granovetter (1978) refers to individuals with very low thresholds for adoption as radicals or instigators, and those with high thresholds as conservatives. He predicts the importance of instigators using binary models, that is, models in which an individual either does or does not adopt by time, t. Speculating relative to ecotoxicology, other reasons for a lag might be the time delay from publication of an innovation until others are able to gather funding, resources, and training, and then begin publishing on the new concept themselves. Also, there might be a delay in convincing conservative opinion leaders to support the concept or methodology. The lag for publications about global distillation and fractionation research seemed to end when regulators began formal efforts to pass the Stockholm Convention for the Control of Persistent Organic Pollutants. That regulatory effort seemed to have produced a quick shift in opinion leaders, those being, regulators controlling research activities through funding. Still other social network features influencing diffusion dynamics will be discussed in Chapter 8, Evidence in Social Networks.

7.3.2 Quantitative Models

Some simple mathematical models can provide more understanding of factors influencing adoption dynamics. The differential models used here come directly from Mahajan and Peterson (1985) who provide succinct and insightful explanations of innovation diffusion models. Although not shown here, integrated models can also be found in Mahajan and Peterson (1985). The most rudimentary diffusion model is analogous to a simple density-dependent population growth model. The diffusion rate $(dN(t)/dt)$ of an innovation into the pool of potential adopters $(\overline{N})$ at time, t, is determined by the difference between the total number of potential adaptors and the number of adaptors present at that particular time $(N(t))$,

$$\frac{dN(t)}{dt} = g(t)\left[\overline{N} - N(t)\right].$$

The $g(t)$ is the diffusion coefficient (probability of adoption at time, t) which depends in this model on the inherent features of the innovation, communication channels, and nature of the involved social system (Mahajan & Peterson, 1985).

An external influence model can be generated by simply replacing $g(t)$ in the general model by a coefficient of external influence (a),

$$\frac{dN(t)}{dt} = a[\overline{N} - N(t)].$$

The associated diffusion of an external diffusion model is determined by external factors such as communication of the innovation by mass media. Pertinent here might be a regulatory edict announced in the *Federal Register* or some other publication in the peer-reviewed literature. The strong external influence means that coefficient (a) remains essentially independent of the number of individuals who have adopted the innovation at any particular time, t. The rate of innovation slows as $N(t)$ approaches $\overline{N}$ but the coefficient remains a constant. This external influence model produces the curve depicted in Fig. 7.3 with a dotted line. This model (Geroski, 2000; Kadushin, 2012) produces a smooth curve that gradually converges at an asymptotic value $(N(t) \to \overline{N}_\infty)$ much like the conventional monoexponential contaminant bioaccumulation model widely used in ecotoxicology. This becomes clear if the integrated external influence diffusion model,

$$N(t) = \overline{N}\left(1 - e^{-at}\right),$$

is compared to the familiar bioaccumulation model,

$$C(t) = C_{Source}\frac{k_u}{k_e}\left(1 - e^{-k_e t}\right),$$

where $C(t) =$ tissue concentration at t, $C =$ source concentration, $k_e =$ the elimination rate constant, and $k_u =$ the uptake clearance rate.

In contrast, the sigmoid curve in Fig. 7.3 reflects a model in which internal communication processes are prominent during adoption. Geroski (2000) gives the example of the different adoption processes for some concepts and technologies. A simple concept might be readily adopted after potential adopters are exposed to it through mass media. In contrast, the adoption of a complex technology might require more person-to-person skill development and word-of-mouth transmission of crucial details. How much person-to-person or word-of-mouth transmission occurs will depend on how many individuals have already adopted the innovation. Diffusion that is dependent on the number of previous adaptors can be modeled with this straightforward differential equation,

$$\frac{dN(t)}{dt} = bN(t)\left[\overline{N} - N(t)\right].$$

This internal influence model results in the sigmoid model shown in Fig. 7.3, which might be fit with a Gompertz, probit, or logistic function (Kadushin, 2012; Mahajan & Peterson, 1985). The index of imitation (b) reflects the intensity of interactions among adopters and potential adopters. As an example, Strang and Meyer (1993) describe socially isolated physicians as adopting the use of a new drug as conforming to an external influence diffusion model, whereas doctors integrated into large networks adopted the new drug according to an internal influence sigmoid model.

"Pure imitation" models (Mahajan & Peterson, 1985) like the above internal influence model might be inadequate in many cases and a mixed-influence model might be more appropriate. A mixed-influence model includes features that are independent and dependent on $N(t)$. A widely applied model of this kind combines the above external and internal influence models,

$$\frac{dN(t)}{dt} = [a + bN(t)]\left[\overline{N} - N(t)\right].$$

The sigmoid cumulative adoption for this model can also be fit with generalized functions like the logistic function (Mahajan & Peterson, 1985). More complex models are possible as described in Granovetter (1978), Mahajan and Peterson (1985), Valente (1996), Geroski (2000), Bettencourt, Cintrón-Arias, Kaiser, and Castillo-Chávez (2006), Friedkin, Proskurnikov, Tempo, and Parsegov (2016), and other sources. They might be required for a variety of reasons or situations. For example, Bettencourt et al. (2006) formulated models that, in addition to the implied "ignorant" and "spreaders" states of individuals in the above models, include a "stifler" state that acts to reduce the rate of diffusion. Details about the heterogeneity or structure of the involved social networks might also be required in some modeled situations. Finally, the models shown above incorporate the assumption of a static system in which the innovation does not change. However, an innovation might be improved or modified by early adopters so as to increase its appeal to later adopters. Mahajan and Peterson (1985) developed this sort of dynamic diffusion model by allowing the pool of potential adopters ($\overline{N}$) to increase over time as a function of a vector of variables reflecting changes in the innovation arena, $f(S(t))$.

$$\frac{dN(t)}{dt} = [a + bN(t)][f(S(t)) - N(t)].$$

Such a model seems appropriate for some groups of ecotoxicologists and ecological risk assessors. Often an innovation is introduced by one set of authors and then refined through a sequence of publications. As the innovation becomes more appealing, the pool of potential adopters increases. Several examples will be discussed in Section 7.5.2.

7.3.3 Strength of Weak Ties

A few points require a bit more detail before we can explore illustrative innovations from ecotoxicology and ecological risk assessment. How an innovation is first introduced into a group was mentioned only very briefly in Chapter 2, Human Reasoning: Everyday Heuristics and Foibles (Section 2.3.2.7) and a bit more detail is warranted at this point. An innovation that solves a problem or answers an unresolved question of a group very often comes from outside of that group. This is accomplished by an individual who actively engages in heterophilic exchange with members of another group. Again, this individual is not often an opinion leader within the group and, due to the dissonance that might be created by the introduction of new ideas from outside of the group, risks being marginalized by opinion leaders highly vested in an existing paradigm. The strength of weak ties is the name given to the common observation that solutions to unsolved problems for a group often are found by such individuals from outside sources. As will be defined more explicitly in Chapter 8, Evidence in Social Networks, a tie is a link connecting two individuals which, in this instance, reflects the transmission of a new concept from one person to another. Weak refers to the relative paucity of such ties between the outside group and the group that receives it. Ties are weaker between groups than within groups. However, these weak ties are crucial to solve outstanding problems within a group, that is, the strength of weak ties is their capacity to bring solutions into a group.

We found a good illustration of a bridge between two groups ("components") of co-authors publishing about the mountain cold-trapping paradigm in 2013 (Fig. 7.4).[6] Most of the co-authorship ties in this network were

---

6. Mountain cold-trapping is "the enrichment of certain persistent organic compounds in various media (e.g., soil, water, snow, or tissues) with increasing elevation up mountain slopes. The mechanism for cold-trapping, as implied in its name, is compound transition out of the atmospheric gaseous phase and into solid phases such as particulates, water droplets (rain and fog), snow, and components of the land surface at the colder temperatures that prevail at higher elevations" (Newman, 2015). See Appendix 1, Ecotoxicology Innovation Survey Methods, for more detail about this and other innovations mentioned in Chapters 7 and Chapter 8.

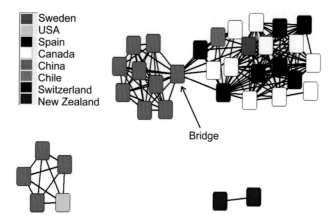

FIGURE 7.4   Co-authorship network for mountain cold-trapping for the year 2013. Each symbol represents a co-author ("node") with lines ("ties") connecting co-authors. Symbol color reflects the country with which an author was affiliated. The country key is placed in the top left corner of the figure. Much of the co-authorship clustering can be explained by mountain ranges of common concern to sets of co-authors. One Chinese co-author formed a crucial bridge between the North American/European and Chinese co-authorship network components.

strongly influenced by the specific mountain ranges being studied. Two New Zealand authors published on the Southern Alps of New Zealand. A cluster of Chinese co-authors studied contaminant movement in the Tibetan Plateau and adjacent mountains, the Himalayas, and mountains around Beijing. A multinational cluster of Europeans and North Americans published on contaminants in mountain ranges of those regions of the world. Illustrating an important tie between two region-based clusters was one investigator from a State Key Laboratory (Chinese Academy of Science) who created a bridge between the large North American/European and Chinese clusters. The resulting tie facilitated the movement of related concepts and methods for studying an emerging issue in Asia.

7.3.4   Temporal Influences

The influence of advances in information technology must be considered when exploring innovation diffusion into ecotoxicology and ecological risk assessment. The ability to communicate and access new evidence and concepts has greatly improved in the period during which ecotoxicology and ecological risk assessment were maturing. The fields of ecotoxicology and ecological risk assessment are also so young that they were still passing through early stages of maturation as various novel concepts and techniques were introduced. Early innovations were introduced as environmental sciences and regulations were just taking shape, whereas other innovations were introduced into a more established science and more formal environmental risk assessment practice.

The *milieu* into which innovations were introduced has changed through time is suggested using 18 conceptual and technical innovations described in Appendix 1, Ecotoxicology Innovation Survey Methods. Plots of the cumulative number of (unique) authors who published on the innovation through time showed a consistent pattern regardless of the innovation. There was a period after first publication of the innovation when very few additional authors were added each year (e.g., Fig. 7.2). After this lag period, the number of new authors added annually increased in a linear fashion for these 18 innovations with no indication of any slackening. Consequently, the duration of the lag period and slope of the linear ascending portion of the curve were picked as convenient metrics of innovation diffusion dynamics of this set of innovations.

A clear linear trend over time is obvious for the duration of lag periods of the 18 innovations (Fig. 7.5, top panel). At one extreme, the lag period for biomagnification lasted for several decades. At the other extreme, the BLM had lag period duration of only a few years. Although no definitive explanation can be given for the trend in lag period shortening through time, the basic external and internal influence models can be used to propose one general explanation. Recollect that the external influence model has no initial lag, whereas the internal influence model has a slow adoption phase until the curve begins to turn sharply upward (Fig. 7.3). The shortening of the innovation adoption lag periods with time could reflect a transition from a predominantly internal influence dynamic to an external influence dynamic. Initially, practicing ecotoxicologists were spread among many universities and departments with no large consolidated groups focused on pollutant effects. One of the early researchers who introduced the biomagnification innovation was George Woodwell, a general ecologist in the Department of Biological Sciences at the State University of New York (Stony Brook). He had very few department colleagues who shared his interests. No prominent journals emphasizing ecotoxicology or environmental pollution existed, and professional societies addressing these themes had yet to emerge. Eldridge Hunt of the California Department of Fish and Game published his report of DDT biomagnification in the quarterly *California Fish and Game*. As time progressed, conditions changed with the emergence of established journals, funding sources, societies, and research groups that focused on pollution. Also, communication of novel scientific ideas became progressively more effective as information technologies improved and publications became easily retrievable electronically. The BLM was introduced by a well-established collaborative team with a history of engagement with environmental regulators. Their initial work was partially funded by a cooperative agreement with EPA, an agency that did not exist when Woodwell published on biomagnification. It was introduced via a series of well-established journals focused on

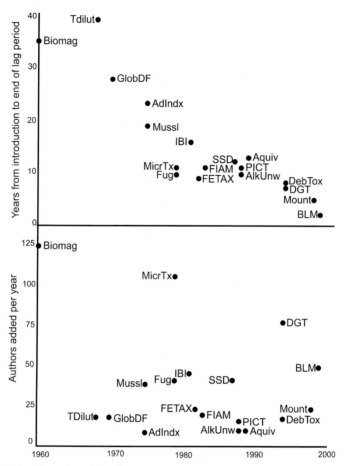

FIGURE 7.5 The estimated lag time (top panel) and also accumulation of authors for each innovation (bottom panel) as a function of calendar year. The mean and 95% highest density interval for the number of authors added per year during the post-lag ascending portion of the curves were 30.57 and 14.86−46.84, respectively. AdIndx = Marking's additive index, AlkUnw = alkaline unwinding assay, Aquiv = aquivalence, Biomag = biomagnification, DebTox = dynamic energy budget-toxicity approach, FETA = frog embryo teratogenesis assay—*Xenopus*, Fug = (ecotoxicological) fugacity model, GlobDF = global distillation and fractionation, IBI = index of biotic integrity, MicrTx = Microtox assay/test, Mount = mountain cold-trapping, Mussl = mussel watch, and Tdilut = trophic dilution. See Appendix 1, Ecotoxicology Innovation Survey Methods, for more explanation and key citations for these innovations.

environmental pollution (i.e., *Environmental Science & Technology* and *Environmental Toxicology and Chemistry*) and quickly gained the support of opinion leaders. As will be discussed in Section 7.5.2.1, rapid acceptance might also have been influenced by it being a refinement to several earlier innovations.

Speculating, word-of-mouth and person-to-person transmission (internal influences) were very important in the innovation diffusion process as ecotoxicology first emerged as a science. As research groups, funding sources, societies, and mass media outlets (established journals) for pollution research emerged and consolidated into an integrated whole, external influences on innovation dynamics are predicted to become increasingly important. If true, the net result would have been the observed transition from a model with a distinct initial lagging tail (Fig. 7.3, internal influence model) to one with almost no lag period (Fig. 7.3, external influence model).

Although the above speculation is plausible, it remains unresolved whether this linear pattern resulted from a slow maturing of the science, advances in information technology, general increase in research collaboration, a combination of these factors, or some other factors. Regardless, the present research community will more quickly publish articles about an innovation than the community did in the 1960s and 1970s. That is to say, good ideas now have a better chance of getting prompt attention. Concomitantly, there is a higher risk of a bad idea gaining too much attention before careful vetting.

There was no obvious temporal trend in the number of authors added each year (Fig. 7.5, bottom panel). Some innovations had a sustained high level of interest in the community (e.g., biomagnification, Microtox assay, DGT) perhaps because of their exceptional usefulness in providing ready answers to applied questions. In contrast, other innovations had shallower trend lines as in the cases of Marking's additive index for toxicant mixtures, the alkaline unwinding assay for DNA damage, and the aquivalence criterion for modeling nonvolatile contaminants.

Two points are suggested from this sampling of innovations. Changes in the fields of ecotoxicology and environmental risk assessment, information technology, and research collaboration could have produced innovation adoption curves that are different from those predicted by the conventional Kuhnian or innovation diffusion curves. However, these are young fields. Conformity to the classic models will likely occur given more time for their dynamics to be fully expressed. Speculating from the existing evidence, the shortening of lag periods might have resulted in a transition from internal influence to external influence dynamics. Regardless, the duration of the lag period has shortened so that an innovation—good or bad—now has a much better chance of gaining early attention than when the fields of ecotoxicology and environmental risk assessment first came into being.

7.4 INFORMATION TRANSMISSION INTEGRITY AND FRUITFULNESS

Information transmission, including transmission of innovations, has two characteristics that need brief mention, integrity and fruitfulness. Explanation will begin with Morgan's (2011) definition of a fact as a unit of "useful and reliable knowledge." Some facts can be paradigms (i.e., a concept or approach that, at a particular moment, provides the most useful and reliable explanation for existing evidence), whereas others might be simply reliable results from surveys and experiments.

According to Morgan (2011), a fact moves with integrity if it remains relatively intact as it is transmitted through a social group or between social groups. Loss of integrity is an undesirable outcome for a fact like survey or experimental evidence although, in some instances pertinent here, it might be desirable for a technique or paradigm. Under the conditions set out in the previous paragraph, some loss of integrity of a technique or paradigm might be desirable if change results in refinement. For instance, the mussel watch innovation instigated by Goldberg (1975b) was framed around one marine taxonomic group ("*Mytilus edulis* and similar species") but this global biomonitoring paradigm was modified through time to create other valuable biomonitoring schemes involving other taxonomic groups, environments, and spatial and temporal scales.

In other cases, loss of fidelity can corrupt an otherwise sound paradigm. As a conceptual example, there are numerous published editorials and diatribes lamenting that the eroded integrity of the *eco*toxicology paradigm (Truhaut, 1977) has created a serious deficiency in the science. (See Section 1.2.2 in Chapter 1, The Emerging Importance of Pollution, for details on Truhaut's definition of ecotoxicology). Given what ecotoxicologists now do, ecotoxicology seems more of an aspirational, than an accurate, moniker. A more accurate name and definition of what most (not all) ecotoxicologists currently do is envirotoxicology, the study of contaminant concentrations and fate in natural environments, and effects to individual organisms inhabiting contaminated environments. As a technical example, the species sensitivity distribution (SSD) paradigm has had some of its integrity eroded relative to its original 1970s context. Although guided by toxicological evidence, it was not originally intended as a scientific paradigm. The original EPA pragmatism that prompted what eventually became the SSD approach was summarized by Stephen et al. (1985):

Because aquatic ecosystems can tolerate some stress and occasional adverse effects, protection of all species at all times and places is not deemed necessary. If acceptable data are available for a large number of appropriate taxa from an appropriate variety of taxonomic and functional groups, a reasonable level of protection will probably be provided if all except a small fraction of the taxa are protected, unless a commercially or recreationally important species is very sensitive. The small fraction is set at 0.05 because other fractions resulted in [water quality] criteria that seemed too high or too low in comparison with the sets of data from which they were calculated. Use of 0.05 to calculate a Final Acute Value does not imply that this percentage of adversely affected taxa should be used to decide in a field situation whether a criterion is too high or too low or just right.

The tone of these and other related statements emerging from early applications of what would eventually become the SSD approach is that of satisficing to accommodate diverse commercial, environmental, economic, human health, and legal concerns. The SSD approach appeared to have met the minimum requirements of parties involved in the regulatory process of reducing contaminant impacts on human health and aquatic biota.

Stephan (2002) reviews the early evolution of the SSD approach to fulfilling a specific regulatory need. At the onset, an ad hoc judgment from available toxicity evidence had been used to derive water quality criteria. As the thinking of regulators advanced, more objectivity was deemed necessary, because as already discussed, universal introspection is error prone and difficult to defend. The most sensitive species approach that used the lowest of all available toxicity metrics often resulted in unrealistically low criteria: that approach did not meet the minimum requirements of the regulated community. Conceptually, it could also be too high if the most sensitive species were not among those tested. In that case, it might also have been compromised in the eyes of agencies obligated to protect fish and wildlife. As described in the above quote, the decision was eventually made to define a regulatory concentration below which only a low percentage of species toxicity metrics occurred. A lognormal distribution was selected for convenience to estimate the concentration corresponding to a certain low percentage of species toxicity metrics. This pragmatic application of satisficing allowed regulatory progress.

As the SSD approach evolved, it began to be interpreted as predicting the concentration below which natural communities will be protected from toxicant impacts. This is ecological nonsense because the species interactions so essential to communities were neglected in these misinterpretations of SSD results. Through further satisficing, the number of required species was reduced from Stephan et al.'s "a large number of appropriate taxa from an appropriate variety of taxonomic and functional groups" to roughly five to eight (Newman et al., 2002; Suter, 2002) although formal analysis suggested that many more were required to be representative of the biota being protected (Newman et al., 2002). The lognormal distribution was generally applied in SSD estimations based on its historical use for this purpose although 27 of 51 data sets assessed by Newman et al. (2002) failed tests of lognormality. Diminished integrity of the SSD concept formulated initially during regulatory satisficing occurred as it evolved.

Morgan (2011) also emphasized the importance of fruitfulness as a fact moves within and among groups. By fruitfulness, she meant how widely the fact or concept moves relative to geographic or disciplinary space. We have already examined fruitfulness in terms of diffusion theory. However, an important point to make from Morgan's particular vantage is that something can spread fruitfully but with low integrity, and vice versa. Certainly, the ecotoxicology paradigm has been fruitful, but many would argue that it has lost some of its integrity relative to the original paradigm proposed by Truhaut (1977).

7.5   INNOVATIONS IN ECOTOXICOLOGY

Specific instances of innovation in ecotoxicology and ecological risk assessment can now be explored with the conceptual tools provided by science historians and sociologists. Some adjustments are required for reasons already mentioned. Both fields are young and the associated innovations have not had enough time to pass through all phases of adoption. Notably, most have yet to approach the saturation or plateau phase (Fig. 7.3). Also, the maturing of both fields has taken place simultaneously with profound advances in information technology and a trend toward more team-oriented research. For these reasons, the innovation results are not yet best fit to conventional models. Simpler models had to be applied until more innovations have run their courses and the influence of advances in information technologies can be incorporated, perhaps with a mixed-influence or dynamic diffusion model.

7.5.1   Fundamental Tension: Standardization and Scientific Evolution

Most readers at this point are likely quite tired of the repetitive use throughout this book of the phrase, ecotoxicology and ecological risk assessment. This admittedly tedious approach was taken to underscore that these distinct fields are sometimes treated as identical and associated activities intermingled in a confusing manner. But, as described in Chapter 4, Pathological Reasoning Within Sciences (Section 4.1.2), these parallel activities have distinct goals and valued qualities. Activities in a science like ecotoxicology achieve the highest explanatory power when they are conducted under the norms and methods set out in Chapter 3, Human Reasoning: Within Scientific Traditions and Rules. Qualities that enhance the utility of a technical endeavor like ecological risk assessment are effectiveness (including cost effectiveness), precision, accuracy, appropriate sensitivity, consistency, clarity of outcome, and ease of application. Some of these technology qualities are also valued in a science; however, some are less important (e.g., ease of application or cost effectiveness) in scientific than in technical endeavors. Meticulous experimental design and logical rigor are weighted much more heavily in a science, whereas they might be lessened somewhat in a technology to accommodate cost effectiveness, consistency, clarity of outcome, or ease of application.

Establishing a clear commitment to either a scientific or technical goal is crucial before conducting or evaluating innovative work in our fields. Neither activity is right or wrong until the activities of one are misconstrued or misjudged using criteria of the other. As pointed out in Chapter 3, Human

Reasoning: Within Scientific Traditions and Rules, ecotoxicologists who apply the SSD method to predict toxicant effects to ecological communities commit a segregation error, that is, a technology designed primarily to address regulatory needs is mistakenly confused with a scientifically sound instrument for assessing the ecological state of an exposed community.

7.5.2   Examples

A few ecotoxicology and ecological risk assessment innovations provide illustrations or insights about the fate of novel ideas or techniques in these fields. The innovation examples explored here were generated during the study described in Appendix 1, Ecotoxicology Innovation Survey Methods.

7.5.2.1   Progression of Innovations Concerning Metal Bioactivity

Kuhnian interpretation of paradigm evolution might at first glance seem pertinent to the linked sequence of innovations associated with understanding metal bioactivity in aquatic environments (Fig. 7.6). Yet, the process that I watched unfold did not have the combative tenor that permeates Kuhn's descriptions of scientific progress. Instead the quality of the associated exchanges and debates was that hoped for in any healthy science.

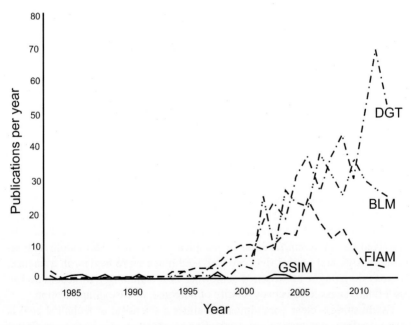

FIGURE 7.6   Progression in the number of publications each year for four innovations contributing to our understanding metal bioavailability in aquatic environments. Except for GSIM, details for these innovations are provided in Appendix 1, Ecotoxicology Innovation Survey Methods.

The progression began with the early adoption of several concepts from classic toxicology. The ionic hypothesis (Mathews, 1904) that the dissolved form of a metal was the most bioavailable form was borrowed. Also taken from general toxicology was the paradigm that differences in metal toxicities were related to differences in metal−ligand binding tendencies (Jones & Vaughin, 1978; Jones, 1939, 1940; Loeb, 1902; McGuigan, 1954; Williams & Turner, 1981). Building upon these premises, Pagenkopf and colleagues (Pagenkopf, 1983; Pagenkopf, Russo, & Thurston, 1974) formulated a conceptual model for the influence of metals on freshwater fish gills. Their Gill Surface Interaction Model (GSIM) focused on acute toxicity to fish and assumed that gill respiratory failure was the crucial consequence of metal-to-gill binding. Their sound thinking based on accepted classic toxicological principles did not result in many publications (Fig. 7.6). Perhaps this was due to the narrow focus on freshwater fish and gill respiratory dysfunction during acute exposures. Next, the Free Ion Activity Model (FIAM) emerged, proposing that the free (aquated) metal ion was the most bioavailable of dissolved metal species.[7] Although related studies had been published in the 1970s (Campbell, 1995), the first clear elucidation of the FIAM seems to have been in the textbook by aquatic chemist, François Morel (1983) during general discussion of metal complexation and speciation. Emphasis in that chapter was speciation in natural waters and how water chemistry influenced metal species activities. Morel explicitly stated that "... the toxicity of many trace metals to planktonic microorganisms is determined by their free ion activities, not their total soluble concentrations." The focus of the FIAM is broader than that of the GSIM, embracing species other than freshwater fish and toxic impacts other than those to fish gills. It underscored the role of the free ion. This closer linkage to aquatic geochemistry made the calculation of potentially bioavailable metal species activities more tractable, including those of free ions, lipophilic organic complexes, and neutral metal complexes (Campbell, 1995). As depicted in Fig. 7.6, this broadening and refining of concepts resulted in a substantial number of FIAM publications. Combined with chemical thermodynamic equilibrium−based computer models, it provided a sound context for examining the influence of pH, salinity, and hardness cations (concentrations of Ca^{2+} and Mg^{2+}) on toxicity, and eventually the bioavailability of metals in sediment interstitial waters. The next refinement was the BLM. As stated in the introduction of Di Toro et al. (2001), the Biotic Ligand Model (BLM) synthesized existing ideas to provide a predictive model for acute metal effects. It combined features of the GSIM and FIAM, generalizing from a gill surface ligand context to include generic

---

7. The terms, free ion and aquated ion are both used to reflect the same moiety. They might be designated M^{2+} and $M(H_2O)_6^{2+}$, respectively. The difference is that the term, aquated ion, reminds us that the ion is surrounded by a hydration sphere of oriented water molecules that can influence interactions.

biotic ligand sites of action. Death was believed to occur when the amount of metal bound to some critical biotic ligand exceeded a threshold concentration. The approach was refined enough at this stage to elicit clear proclamations like "The scientific underpinnings of the present BLM approach are sound" (Niyogi & Wood, 2004). Once integrated with chemical equilibrium computer models, it became the basis for numerous water quality decisions. In short time (Fig. 7.6), the annual numbers of new BLM publications surpassed those for the FIAM. With the foundation laid for assessing metal effects, computational and field methods advanced for generating relevant data. Convenient chemical equilibrium computer programs (e.g., Visual MINTEQ) replaced more cumbersome and outdated programs, allowing estimation of metal species including metal complexes with natural dissolved organic ligands. The rapid generation of "labile" metal data with DGT passive sampling technology (Allan et al., 2007; Davison & Zhang, 1994; Davison, Grime, Morgan, & Clarke, 1991; Li, Zhao, Teasdale, John, & Wang, 2005; Zhang & Davison, 1995) is one important example (Fig. 7.6).

Stepping back to understand the general pattern for this sequence, the central paradigm that dissolved metal coordination chemistry determines bioactivity was adeptly borrowed from classic toxicology. It remained relatively intact in ecotoxicology during the integration process, diffusing fruitfully and with high integrity. Speculating from the lag times in plots of the cumulative number of authors through time, each successive innovation appeared to hasten the acceptance of the next. The GSIM was explicitly introduced in 1983 but still had not emerged from its lag period by 2013, so the best estimate for its lag period is greater than three decades. In chronological order, the lag times then shortened progressively from 11 (FIAM), to 7 (DGT), to 2 years (BLM) (Fig. 7.5).[8]

During the integration phase, satellite concepts and technologies were refined to directly address effects to aquatic biota. This entailed no more than a healthy amount of debate. No Kuhnian rancor emerged because no one's professional interests seemed seriously threatened by the incremental innovations.

Each innovation had qualities that Rogers (1995) had identified as promoting adoption. The perceived advantages were obvious for each innovation. The initial premise that bioactivity can be predicted from total dissolved metal concentration was steadily improved upon by each of these innovations. It did so in a manner consistent with paradigms that had emerged in other fields. It provided sound explanations for the influences of

---

8. The short lag period of mountain cold-trapping compared to global distillation/fractionation also supports this preconditioning feature of innovation adoption. Mountain cold-trapping and global distillation/fraction are based on the same concepts except one involves changes with altitude and the other changes with latitude. Adoption of global distillation/fractionation would foster later adoption of mountain cold-trapping.

alkalinity, hardness metals, pH, and salinity on the toxic impact of dissolved metals. Although initially difficult to assess using anodic stripping voltammetry, prediction of metal species activities during the period of adoption became increasing easier with improved specific ion electrodes (especially the cupric ion electrode) and chemical equilibrium computer models. The risk associated with accepting each innovation on a trial basis decreased as their applications became easier. The advantages of each innovation became increasingly apparent as the number of published applications increased. The advantages to regulators were so apparent that the approach gradually became a common theme in water quality criteria documents. None of the impediments outlined by Klein and Knight (2005) stood in the way of adoption.

So, the process began with a paradigm with a firm scientific foundation and possessing qualities that foster rapid adoption. Innovations were incrementally refined and provided ancillary support to this paradigm. Equally important from a scientific context, the refined paradigm was explicit and quantitative: it could be rigorously tested and modified by application of scientifically accepted methods.

As Kuhn observed, all paradigms are eventually replaced by superior ones. There is no reason to assume that the BLM will be an exception. As pointed out in some of the first BLM publications (Niyogi & Wood, 2004), the context of the current BLM is acute exposure. The BLM might not be sufficient for chronic exposures because binding sites change during biological acclimation to low metal concentrations. Speculating, the past emphasis on equilibrium models was one of convenience and likely a more realistic kinetics approach will be integrated into our thinking about metal bioactivity. Also, the emphasis on dissolved metal ignores the very substantial exposure to metals in ingested materials such as sediment, suspended solids, and biological tissues. Different mechanisms have evolved in the digestive systems of diverse species that will require accommodation in any conceptual model of environmental metal bioavailability and bioactivity.

7.5.2.2 Innovations Differing in Ease of Application, Simplicity, and Relevance

Conventional thinking about the advance of science is that an innovation that provides a better explanation of existing evidence will be adopted as the best current explanation. According to innovation diffusion theory, the speed of adoption will be accelerated by its perceived advantage relative to the current explanation, its compatibility with existing theories and evidence in the field, conceptual simplicity, ease of implementation, trialability, and its perceived advantages by potential adopters. Also, simple technologies are adopted more readily than complex ones.

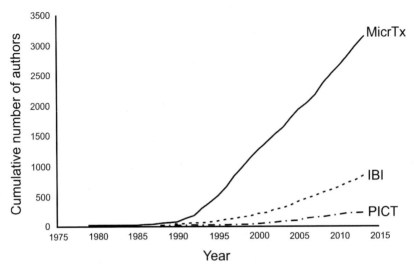

FIGURE 7.7   The cumulative number of authors for three innovations that differ in their perceived ease of application and conceptual simplicity. IBI = index of biotic integrity and MicrTx = Microtox assay/test. Details for these innovations are provided in Appendix 1, Ecotoxicology Innovation Survey Methods.

A comparison of three innovations, the MicroTox test, Index of Biotic Integrity (IBI), and Pollution-induced Community Tolerance (PICT) concept, can be used as an illustration that factors in addition to scientific soundness influence the success of a new idea or technique in ecotoxicology (Fig. 7.7). Although the MicroTox test provides much less insight than do the IBI and PICT about the impact of toxicants to ecological communities, the number of authors who published articles using the MicroTox test increased much more rapidly than those for the IBI and PICT. This initially counterintuitive observation makes sense if the three are compared simply as innovations, not as tools with varying degrees of explanatory or predictive power. The MicroTox test was introduced as "a simple rapid method" that gave results comparable to fish toxicity tests (Bulich, 1979). Johnson (2005) described it as a tool for "minimalistic microscale toxicity testing." Its intended use was not to imply ecotoxicological consequences, and its virtues are those of an innovation that would be rapidly adopted and implemented.

In contrast to the MicroTox test, the PICT directly measures an emergent quality of microbial communities exposed to contaminants. It is based on the premise that tolerance to toxicant effects on some community quality such as photosynthesis results from shifts in species composition of the community toward the most tolerant species, and genetic adaptation of the exposed populations. Detection of induced tolerance was viewed as "direct evidence that a community is affected by toxicants in an ecosystem" (Blanck, Wängberg, & Molander, 1988). Incorporating both community and population processes made it very ecologically relevant; yet, the concept attracted very few authors (Fig. 7.7). Several basic qualities of an innovation seem to have contributed to this low number of authors. The mechanisms for community tolerance are more complicated than the simple one behind the MicroTox test. If enhanced tolerance was evident in an exposed microfloral community, was it due to shifts in community composition, genetic adaptation, or both? Is PICT to be expected in all, or just some, kinds of microfloral communities? Although it was consistent with existing theory, its advantage was unconvincing to those seeking clear evidence of toxicant harm to valued components of ecosystems. If the shift in tolerance occurred, was that not a sign that the impact of the toxicant exposure was ameliorated? Does the presence of PICT mean there was or was not an undesirable impact? Finally, it was difficult to convince many, including regulators and the public, that shifts in microfloral communities reflected unacceptable damage to valued ecological entities such as the fish assemblage of a water body.

The IBI seems to be positioned between the MicroTox test and PICT. It is more difficult to conduct an IBI-based study than a MicroTox test, but many aquatic and fisheries scientists have the expertise to conduct an IBI survey. Various components of the IBI such as species richness, diversity, and other attributes are well accepted metrics of community structure (Karr, 1981). Its trialability is high because the IBI is derived from information normally taken during routine fish surveys of flowing water bodies. Linkage of the results to the health of a highly valued entity, the stream fish assemblage, provided more convincing evidence of undesirable community impact than did results of the MicroTox test or PICT.

7.6 CONCLUSIONS

The question was posed in the beginning of this chapter about what factors other than scientific soundness influence the fate of an innovation introduced into ecotoxicology or ecological risk assessment. Some preliminary answers can now be provided.

Although it might come as a surprise, Kuhn's paradigm revolution framework seems only moderately helpful for answering this question. With the two exceptions already discussed, most innovations to date in ecotoxicology have not been associated with fundamental paradigm shifts. And the only future exceptions might be the global distillation/fractionation paradigm and perhaps the emerging environmental amplification context (MacDonald et al., 2002) for studying POP bioconcentration, bioaccumulation, and biomagnification. Most instances of apparent conflict involving innovations such as those examined here likely reflect professional uncivil behavior, and not a normal conflict expected between innovators and vested parties during paradigm replacement. A Kuhnian crisis would likely occur for more fundamental paradigms in ecotoxicology. Considerable conflict would be expected if the seldom-articulated paradigm was questioned that "ecological consequences of environmental exposures to chemicals can be reliably predicted from individual-based testing." Substantial conflict would also be expected if assertions were made to replace Fisherian frequentist statistical methods used in ecotoxicology with an alternative such as Neyman–Pearson, neoFisherian, information-theoretic, or Bayesian methods.

The trends noted for ecotoxicological innovations seem to result from differences known to influence general innovation diffusion. Examples include the ecotoxicology concept itself, DGT, BLM, FIAM, global distillation/fractionation, GSIM, saprobic index, mountain cold-trapping, and the SSD innovations. Many innovations in ecotoxicology are refined as they diffuse through the pool of potentially adopting scientists and risk assessors. The GSIM-FIAM-BLM sequence and the mussel watch biomonitoring innovation are examples of beneficial changes to an innovation during diffusion.

The movement of novel concepts and techniques into ecotoxicology and ecological risk assessment occurred simultaneously with the maturation of these fields, a shift toward more collaborative research, and extraordinary advances in information technology. Evidence of the consequences is suggested by trends in lag times for 18 innovations (Fig. 7.5). The linear decrease in lag times from the 1960s to 2013, and linear shape of the post-lag portion of the curves suggests that external influences became increasingly important with time. The fields matured and more associated journals (the mass media of sciences) became available. Also the ability to search, identify, and consider an innovation for adoption advanced dramatically with the advent of electronic search tools and conversion of publications to searchable electronic formats. Speculating further, the increasingly common conduct of research by large teams might also have accelerated innovation adoption. This has created a great advantage but, because bad innovations can now also enter quickly into the scientific community, it creates an obligation to very carefully scrutinize novel concepts and technologies as they appear.

REFERENCES

Allan, I. J., Knutsson, J., Guigues, N., Mills, G. A., Fouillac, A. M., & Greenwood, R. (2007). Evaluation of the Chemcatcher and DGT passive samplers for monitoring metals with high fluctuating water concentrations. *Journal of Environmental Monitoring, 9*, 672–681.

Berkson, J. (1951). Why I prefer logit to probits. *Biometrics, 7*, 327–339.

Berkson, J. (1953). A statistically precise and relatively simple method of estimating the bioassay with quantal response, based on the logistic function. *Journal of the American Statistical Association, 48*, 565–599.

Bettencourt, L. M. A., Cintrón-Arias., Kaiser, D. I., & Castillo-Chávez, C. (2006). The power of a good idea: Quantitative modeling of the spread of ideas from epidemiological models. *Physica, 364*, 513–536.

Blanck, H., Wängberg, S.- Å., & Molander, S. (1988). Pollution-induced community tolerance— A new ecotoxicological tool. In J. Cairns, Jr., & J. R. Pratt (Eds.), *Functional testing of aquatic biota for estimating hazards of chemicals, ASTM STP* (988, pp. 219–230). Philadelphia: American Society for Testing and Materials.

Bliss, C. I. (1935). The calculation of the dosage-mortality curve. *Annals of Applied Biology, 22,* 134–307.

Bulich, A. A. (1979). Use of luminescent bacteria for determining toxicity in aquatic environments. In L. L. Marking, & R. A. Kimerle (Eds.), *Aquatic toxicology* (pp. 98–106). Philadelphia: American Society for Testing and Materials, ASTM STP 667.

Campanario, J. M. (1993). Consolation for the scientist: Sometimes it is hard to publish papers that are later highly-cited. *Social Studies of Science, 23,* 342–362.

Campbell, P. G. C. (1995). Interactions between trace metals and aquatic organisms: A critique of the free-ion activity model. In A. Tessier, & D. R. Turner (Eds.), *Metal speciation and bioavailability in aquatic systems* (pp. 45–102). Chichester: John Wiley & Sons Ltd.

Chamberlin, T. C. (1897). The method of multiple working hypotheses. *Journal of Geology, 5,* 837–848.

Chutter, F. M. (1972). An empirical biotic index of the quality of water in South African streams and rivers. *Water Research, 6,* 19–30.

Crane, D. (1972). *Invisible colleges. Diffusion of knowledge in scientific communities.* Chicago: University of Chicago Press.

Davison, W., Grime, G. W., Morgan, J. A. W., & Clarke, K. (1991). Distribution of dissolved iron in sediment pore waters at submillimetre resolution. *Nature, 352,* 323–325.

Davison, W., & Zhang, H. (1994). In situ speciation measurements of trace components in natural waters using thin-film gels. *Nature, 367,* 546–548.

Duncan, S. S. (1974). The isolation of scientific discovery: Indifference and resistance to a new idea. *Science Studies, 4,* 109–134.

Finney, D. J. (1947). *Probit analysis. A statistical treatment of the sigmoid response curve.* Cambridge: Cambridge University Press.

Fisher, R. A. (1918). The correlation between relatives on the supposition of Mendelian inheritance. *Philosophical Transactions of the Royal Society of Edinburgh, 52,* 399–433.

Friedkin, N. E., Proskurnikov, A. V., Tempo, R., & Parsegov, S. E. (2016). Network science on belief system dynamics under logic constraints. *Science, 354,* 321–326.

Fuchs, S. (1993). A sociological theory of scientific change. *Social Forces, 71,* 933–953.

Gaddum, J. H. (1953). Bioassays and mathematics. *Pharmacological Reviews, 5,* 87–134.

Geroski, P. A. (2000). Models of technology diffusion. *Research Policy, 29,* 603–625.

Goldberg, E. D. (1975a). Synthetic organohalides in the sea. *Proceeding of the Royal Society (London) B, 189,* 277–289.

Goldberg, E. D. (1975b). The mussel watch—A first step in global marine monitoring. *Marine Pollution Bulletin, 6,* 111.

Granovetter, M. (1978). Threshold models of collective behavior. *American Journal of Sociology, 83,* 1420–1443.

Guanzhong, L. (1995). Three kingdoms, volume 1 (M. Roberts Trans.) Beijing: Foreign Languages Press (Original work published c. 1321).

Hering, D., Moog, O., Sandin, L., & Verdonschot, P. F. M. (2004). Overview and application of the AQEM assessment system. *Hydrobiologia, 516,* 1–20.

Johnson, B. T. (2005). 1. Microtox® acute toxicity test. In C. Blaise, & J.-F. Férard (Eds.), *Small-scale freshwater toxicity investigations* (Vol. 1, pp. 69–105). Dordrecht: Springer Netherlands.

Jones, J. R. E. (1939). The relation between the electrolytic solution pressure of the metals and their toxicity to the stickleback (Gasterosteus aculeatus). *Journal of Experimental Biology*, *16*, 425–437.

Jones, J. R. E. (1940). A further study of the relation between toxicity and solution pressure, with Polycelis nigra as test animal. *Journal of Experimental Biology*, *17*, 408–415.

Jones, M. R., & Vaughin, W. K. (1978). HSAB theory and acute metal ion toxicity and detoxification processes. *Journal of Inorganic and Nuclear Chemistry*, *40*, 2081–2088.

Kadushin, C. (2012). *Understanding social networks. Theories, concepts, and findings*. Oxford: Oxford University Press.

Karr, J. R. (1981). Assessment of biotic integrity using fish communities. *Fisheries*, *6*, 21–27.

Ke, Q., Ferrara, E., Radicchi, F., & Flammini, A. (2014). Defining and identifying sleeping beauties in science. *Proceedings of the National Academy of Science of the United States of America*, *112*, 7426–7431.

Kirton, M. (1976). Adaptors and innovators: A description and measure. *Journal of Applied Psychology*, *61*, 622–629.

Klein, K. J., & Knight, A. P. (2005). Innovation implementation. *Current Directions in Psychological Science*, *14*, 243–246.

Kolkwitz, R., & Marsson, M. (1908). öekologie der Saprobien. 'Uber die Beziehungen der Wasserorganismen zur Umwelt. *Berichte der Deutschen Botanischen Gesellschaft*, *26*, 505–519.

Kolkwitz, R., & Marsson, M. (1909). öekologie der tierusche Saprobien. Beiträge zur Lehre von der biologische Gewässerbeurteilung. *International Review of Hydrobiology*, *2*, 126–152.

Kuhn, T. S. (1970). *The structure of scientific revolutions* (2nd Ed). Chicago: University of Chicago.

Kuhn, T. S. (1977). *The essential tension. Selected studies in scientific tradition and change*. Chicago: University of Chicago Press.

Kuhn, T. S. (2000). *The road since structure*. Chicago: University of Chicago Press.

Li, W., Zhao, H., Teasdale, P. R., John, R., & Wang, F. (2005). Metal speciation measurement by diffusive gradients in thin films technique with different phases. *Analytica Chimica Acta*, *533*, 193–202.

Loeb, J. (1902). Studies on the physiological effects of the valency and possibly the electric charges of ions. I. The toxic and antitoxic effects of ions as a function of their valency and possibly their electric charge. *American Journal of Physiology*, *6*, 411–433.

Loh, C. (2011). Hong Kong—Mainland innovations in environmental protection since 1980. *Asian Survey*, *51*, 610–632.

MacDonald, R., Mackay, D., & Hickie, B. (2002). Contaminant amplification in the environment. *Environmental Science & Technology*, *36*, 457A–462A.

Mahajan, V., & Peterson, R. A. (1985). *Models for innovation diffusion*. Newbury Park, CA: SAGE Publications, Inc.

Majchrzak, A., Cooper, L. P., & Neece, O. E. (2004). Knowledge reuse for innovation. *Management Science*, *50*, 174–188.

McGuigan, H. (1954). The relation between the decomposition-tension of salts and their antofermentative properties. *American Journal of Physiology*, *10*, 444–451.

Morel, F. M. M. (1983). *Principles of aquatic chemistry*. New York: John Wiley & Sons.

Morgan, M. S. (2011). Travelling facts. In P. Howlett, & M. S. Morgan (Eds.), *How well do facts travel?* (pp. 3–39). Cambridge: Cambridge University Press.

Mulkay, M. (1974). Conceptual displacement and migration in science. A prefatory paper. *Science Studies, 4*, 205−234.

Newman, M. C. (2015). *Fundamentals of ecotoxicology. The science of pollution* (4th Ed). Boca Raton: CRC/Taylor and Francis.

Newman, M. C., Ownby, D. R., Mézin, L. C. A., Powell, D. C., Christensen, T. R. L., Lerberg, S. B., ... Padma, T. V. (2002). Species sensitivity distributions in ecological risk assessment: Distributional assumptions, alternate bootstrap techniques, and estimation of adequate number of species. In L. Posthuma, G. W. Suter, II, & T. P. Traas (Eds.), *Species sensitivity distributions in ecotoxicology* (pp. 119−132). Boca Raton: Lewis Publishers/CRC Press.

Nissani, M. (1995). The plight of the obscure innovator in science: A few reflections on Campanario's note. *Social Studies of Science, 25*, 165−183.

Niyogi, S., & Wood, C. M. (2004). Biotic ligand model, a flexible tool for developing site-specific water quality guidelines for metals. *Environmental Science & Technology, 38*, 6177−6192.

Pagenkopf, G. K. (1983). Gill surface interaction model for trace-metal toxicity to fishes: Role of complexation, pH, and water hardness. *Environmental Science & Technology, 17*, 342−347.

Pagenkopf, G. K., Russo, R. C., & Thurston, R. V. (1974). Effect of complexation on toxicity of copper to fishes. *Journal of the Fisheries Research Board of Canada, 31*, 462−465.

Pantle, R., & Buck, H. (1955). Die biologische Überwachung der Gewässer und die Darstellung der Ergebnisse. *Gas-und Wasserfach, 96*, 604−618.

Planck, M. K. (1949). In F. Gaynor. (Ed.), *Scientific autobiography and other papers*. New York: Philosophical Library.

Popp, D. (2010). Exploring links between innovation and diffusion: Adoption of NOx control technologies at US coal-fired power plants. *Environmental and Resource Economics, 45*, 319−352.

Van Raan, A. F. J. (2004). Sleeping beauties in science. *Scientometrics, 59*, 461−466.

Rogers, E. M. (1995). *Diffusion of innovation* (4th ed.). New York: The Free Press.

Ryan, B., & Gross, N. C. (1943). The diffusion of hybrid seed corn in two Iowa communities. *Rural Sociology, 8*, 15−24.

Rzhetsky, A., Foster, J. B., Foster, I. T., & Evans, J. A. (2015). Choosing experiments to accelerate collective discovery. *Proceedings of the National Academy of Science of the United States of America, 112*, 14569−14574.

Sládeček, V., & Tuček, F. (1975). Relation of the saprobic index to BOD_5. *Water Research, 9*, 791−794.

Stephan, C. E. (2002). Use of species sensitivity distributions in the derivation of water quality criteria for aquatic life by the U.S. Environmental Protection Agency. In L. Posthuma, G. W. Suter, II, & T. P. Traas (Eds.), *Species sensitivity distributions in ecotoxicology* (pp. 211−220). Boca Raton: Lewis Publishers/CRC Press.

Strang, D., & Meyer, J. W. (1993). Institutional conditions for diffusion. *Theory and Society, 22*, 487−511.

Stephen, C. E., Mount, D. I., Hansen, D. J., Gentile, J. R., Chapman, G. A., & Brungs, W. A. (1985). *Guidelines for Deriving Numerical National Water Quality Criteria for the Protection of Aquatic Organisms and Their Uses, PB85-227049*. Washington: US Environmental Protection Agency.

Suter, G. W., II (2002). North American history of species sensitivity distributions. In L. Posthuma, G. W. Suter, II, & T. P. Traas (Eds.), *Species sensitivity distributions in ecotoxicology* (pp. 11−17). Boca Raton: Lewis Publishers/CRC Press.

Di Toro, D. M., Allen, H. E., Bergman, H. L., Meyer, J. S., Paquin, P. R., & Santore, R. C. (2001). Biotic ligand model of the acute toxicity of metals. I. Technical basis. *Environmental Toxicology and Chemistry, 20,* 2383–2396.

Truhaut, R. (1977). Ecotoxicology: Objectives, principles and perspectives. *Ecotoxicology and Environmental Safety, 1,* 151–173.

Uzzi, B., Mukherjee, S., Stringer, M., & Jones, B. (2013). Atypical combinations and scientific impact. *Science, 342,* 468–472.

Valente, T. W. (1996). Social network thresholds in the diffusion of innovations. *Social Networks, 18,* 69–89.

Valente, T. W., & Roger, E. M. (1995). The origins and development of the diffusion of innovations paradigm as an example of scientific growth. *Science Communication, 16,* 242–273.

Wejnert, B. (2002). Integrating models of diffusion of innovations: A conceptual framework. *Annual Review of Sociology, 28,* 297–326.

Williams, M. W., & Turner, J. E. (1981). Comments on softness parameters and metal toxicity. *Journal of Inorganic and Nuclear Chemistry, 43,* 1689–1691.

Wray, K. B. (2003). Is science really a young man's game? *Social Studies of Science, 33,* 137–149.

Zhang, H., & Davison, W. (1995). Performance characteristics of diffusion gradients in thin films for the in situ measurement of trace metals in aqueous solution. *Analytical Chemistry, 67,* 3391–3400.

Zuckerman, H. (1996). *Scientific elite: Nobel laureates in the United States.* New Brunswick, NJ: Transaction Publications.

Chapter 8

Evidence in Social Networks

8.1 THE SOCIAL SCIENCE OF ECOTOXICOLOGY

Scientific collaboration is a social process and probably there are as many reasons for researchers to collaborate as there are reasons for people to communicate.

Katz (1994)

In lockstep with advances in information technology have come equally remarkable advances in social media. Social networking sites have changed most aspects of our lives from finding a plumber (e.g., Angie's list), soulmate (e.g., eHarmony), or business partner (e.g., LinkedIn) to shaping political outcomes (e.g., Facebook and Twitter use during the Arab spring and Twitter use by 2016 presidential candidate, Donald Trump). ResearchGate, a social network service for scientists, opened in 2008 and now includes information about more than 13 million researchers.

Recollect that previous chapters explored logical errors and ways of avoiding them, sound and unsound scientific approaches, best practices for statistical inference, and the manner by which innovations diffuse into sciences. Given the directly relevant insights that emerged from these earlier chapters, a no-nonsense reader might question whether any further insight can be gained by examining emergent features of social networks of scientists. Would not any additional insights simply reflect tertiary nuances? Although we might find such insights fascinating as social creatures, would not it be a better use of time to focus on primary determinants of scientific productivity and research quality?

An initial stab at answering these questions can be made with data for 80 ecotoxicologists taken from ResearchGate (Fig. 8.1, Appendix 2). Consistent with Hirsch (2007), plotting h-indices (a measure of an individual's productivity and citation in their literature) against the number of years since an individual's first publication (until 2017) suggests that practicing ecotoxicologists' productivity and associated citation impact increases with time. However, substantial differences in this increase emerge through time with some ecotoxicologists falling far below and others well above average model predictions. As examples, individuals A−F reflect the broad scatter among middle- to late-career ecotoxicologists.

The Nature and Use of Ecotoxicological Evidence. DOI: https://doi.org/10.1016/B978-0-12-809642-0.00008-X

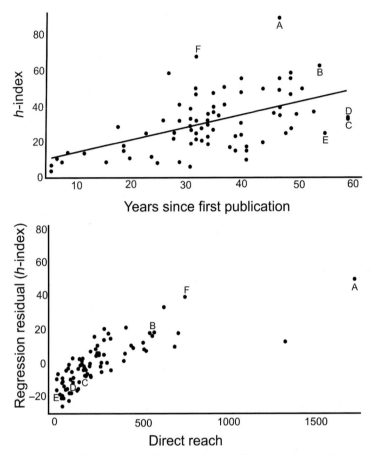

FIGURE 8.1   ResearchGate analysis of network data for 80 ecotoxicologists. The top panel depicts the *h*-index for each individual versus the number of years since their first publication. A linear model was applied to be consistent with Hirsch (2007) who advocated for the wide use of the *h*-index. The regression residuals from this linear model were then plotted in the bottom panel against the ResearchGate Direct reach metric. To maintain anonymity, the labels A−F are applied instead of names for outliers discussed in the text.

A chief contributor of much of the unexplained differences is suggested when the regression residuals from a simple linear model of these data (Fig. 8.1, top panel) are plotted against the ResearchGate Direct Reach metric (bottom panel). The Direct Reach metric is the sum of the number of co-authors, collaborators, and ResearchGate publication and study followers to whom an ecotoxicologist is directly linked, that is, their number of direct network links to others. The six outliers (A–F) all conform to a general trend. Individuals A, B, and F whose *h*-indices were higher than model predictions had careers involving more networking than did individuals

C, D, and E whose h-indices were lower than predicted. Simply stated, individuals with higher than predicted h-indices had the largest network of direct collaborators, co-authors, and followers.

The influences of time in the field and number of professional network connections were roughly similar in magnitude. Individuals who had been publishing for 30 or more years had h-indices ranging widely from 6 to 90. The magnitude of the differences in the Direct Reach metric for individuals A to F was nearly as large (roughly 60 h-index units) as those associated with time in career. The initial conclusion from this survey is that networking has a nontrivial influence on one's impact as reflected by the h-index. Understanding professional networks might provide worthwhile insights about differences in ecotoxicologists' career paths.[1] Having such insights early in one's career might encourage judicious career planning.

Beyond the direct benefits of networking to individual scientists are the broader benefits to society. Jha and Welch (2010) point out that more collaboration is now occurring in response to society's mounting insistence for "a gradual transformation, redefinition and expansion of the practice of academic science and engineering toward broader applicability and transferability of knowledge." Well-executed collaborations increase the frequency of meaningful interactions among scientists (intensity), the range of potential research contents (substance), and diversity of research vantages or purposes (heterogeneity). Research collaborations that enhance intensity, substance, and heterogeneity improve knowledge generation. Multidisciplinary research draws from a wide intellectual pool of ideas and techniques, and is likely to be read more widely by individuals from diverse disciplines. The result is that the associated knowledge is "circulated, validated, and enriched by contact with new research and social circles" (Rodriguez & Pepe, 2008). As a final societal benefit, knowledge and talents emerging from international collaborative teams are more likely to be available for solving transnational problems and addressing the needs in less developed countries that traditionally receive less attention than warranted.

8.2 SOCIAL NETWORK ANALYSIS

Analysis of direct and indirect ties ... revealed that a tie with one or more of the highly productive scientists brought other scientists of lesser productivity into a large network of influence and communication.

Crane (1969)

---

1. Fairness requires the point to be made that success is measured in many ways other than h-indices. Individuals C and D have h-indices below the regression line, yet both are very highly regarded for their innovative research. Individual E received numerous, well-deserved awards for his lifelong contributions to moving the field from its early state to its present healthy condition.

8.2.1   Network Types, Qualities, and Metrics

... social networks affect perceptions, beliefs, and actions through a variety of structural mechanisms that are socially constructed by relations among entities. Direct contacts and more intensive interactions dispose entities to better information, greater awareness, and higher susceptibility to influencing or being influenced by others.

Knoke and Yang (2008)

To this point, most of our discussions have revolved around the individual (also referred to as actor, ego, node, or vertex) or some relation (also referred to as ties, arcs, or edges) between two (dyad) or a few individuals.[2] Discussions of processes emerging in larger groups were laid out only in general terms in previous chapters. Here, a brief treatment of networks of three (triad) or more individuals will be explored to create a fuller appreciation of the social *milieu* in which ecotoxicologists and ecological risk assessors ply their trades. First a basic description of kinds of networks will be provided. Then network qualities and related metrics will be explored, avoiding unnecessarily complex metric computations and matrix methods. The price of limiting the mathematics in this chapter is the inclusion of more figures than typical to illustrate concepts and metrics. Once the basics of social networks are outlined, the reader should have sufficient grasp of concepts and quantitative tools to explore some actual networks of environmental professionals. A reader desiring more quantitative treatment of social networks is encouraged to carefully read Borgatti, Everett, and Johnson (2013) or Knoke and Yang (2008). An excellent treatment of general network qualities and metrics is also provided by Boccaletti, Latora, Moreno, Chavez, and Hwang (2006).

Nodes and connecting ties in social networks can represent a variety of relationships. They could indicate relational events (perhaps information exchange), cognitive states (perhaps an evidence-based decision or judgment), or relational roles (perhaps co-authorship affiliation) among actors (as per Borgatti et al., 2013). Although the mathematical definition of a network given above could include a dyad of two nodes, the simplest network considered in this chapter will be that formed by a triad (Fig. 8.2). Dyads were discussed earlier in Chapter 6, so discussion here is unnecessary. Many treatments of social networks begin with triads because, as Kadushin (2012) points out, the relative connectiveness between individuals only manifests with three or more actors. The third actor increases the complexity of social

---

2. In the context of social networks, the term "relation" is defined explicitly by Knoke and Yang (2008) as "a specific kind of contact, connection, or ties between pairs of actors, or dyads." Although scientific collaboration in the form of co-authorship relations will be the primary focus here, other types are clearly relevant including such relations as ecological risk assessment team reporting, technology transfer, or trust in opinions/judgments.

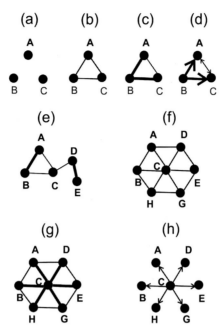

FIGURE 8.2 Simple social networks beginning with undirected/unweighted (second triad from the left in top row), undirected/weighted (third triad from the left in top row), and directed/weighted (fourth triad from the left in top row) examples. The five-member network in the second row is an undirected/weighted network with actor C acting as a bridge between two strongly interacting dyads (actors A and B, and actors D and E). The three hexagonal networks depict differences in exchange in networks with identical numbers of actors.

dynamics by, for example, becoming a mediator, broker, moderator, or tie-breaker. The series of four networks drawn in the top row of Fig. 8.2 suggests how quickly dynamics become complex even for triads. As an extreme case, the individuals A, B, and C do not interact at all in the leftmost triad (a) as indicated by the absence of any ties connecting them: no relationship exists among the actors except the uninteresting one that they are all members of a set. Notice in the next triad to the right that there is no indication of the direction of exchange or weight of ties. Depiction as an undirected/ unweighted network might be appropriate if ties simply indicate co-authorship linkage. In some cases, indication of direction and strength of ties might be warranted for a quality such as opinion/advice exchange or co-authorship networks in which a subset of authors might have published many times together. The next triad to the right (c) has differences in tie weights as indicated by the thickness of tie lines. In that triad, the tie between actors A and C is much weaker than that between actors A and B. Perhaps B published several papers with A and C, but A and C have only published one paper together. Some network ties are directional as in the case of the communication of a new piece of information from one individual to another. As depicted by arrows in the rightmost triad (d), both strength and direction of ties are relevant in such networks. Perhaps actor B in the rightmost triad (d) had the major responsibility of providing information to other members of the triad.

Complexity is quickly added to social exchange if two additional actors are included (Fig. 8.2e). Notice that actor C, although not possessing strong ties to any other actor, still influences the network by acting as a bridge connecting dyad A−B to dyad D−E. The reader might recollect that Fig. 7.4 of the co-authorship network for mountain cold-trapping illustrated such a bridge between groups of Chinese authors and North American/European authors. Without actor C, the network would have two disconnected components.[3]

In the context of social networks, a pendant is an actor (like actor E) who has ties to only one other actor (actor D) who, in turn, has ties with others in the network.[4]

Networks can have different qualities even if they have the same number of actors. The three hexagonal networks in Fig. 8.2(f)-(h) have seven actors but their ties differ in important ways. The one in the middle row (f) has equal ties between all actors. As an example of a network of this nature, an ideally functioning scientific team would have all actors sharing information equally. The leftmost hexagonal network in the bottom row (g) reflects the strongest ties between actor C and the other individual actors. Perhaps actor C is the leader of a research team with other members being located at separate analytical service firms. The final hexagonal network (h) indicates that information only goes outward from actor C with no exchange between the other actors. This would be typical of a "tight rein" or "top-down" management style by actor C that would seem less appropriate for leading a scientific or regulatory team comprised of highly motivated and educated members. However, there are instances in which conventional expectations that scientific teams will share information equitably are not realized. A primary author of a "big science" research publication might be required to

---

3. Component is a specific term in network studies. It is easily understood for undirected networks. Nodes are in different components if they cannot reach each other by any path. There are three components in Fig. 7.4, for example. Determining the number of components can be difficult for directed networks. This can be illustrated with the bottom rightmost hexagonal network in **Fig. 8.2**h. Actors A, B, D, E, G, and H cannot reach each other, so they are separate components. Thankfully, components of the undirected co-authorship networks emphasized in this chapter are easily identified.

4. Furthermore, an actor that exists without ties to any other actor is called an isolate. As a relevant example, five publications on a topic might exist with one publication having been written by a single author who never published with anyone else on the topic. That author would be an isolate.

apply the intermediate strategy depicted in the rightmost hexagonal network in the bottom row (g).[5]

Two issues, evolution of networks and ability to associate important attributes to network nodes, need to be touched upon before ending this broad introduction. These features will be illustrated with Figs. 7.4 and 8.3, which show the co-authorship networks for mountain cold-trapping during its initial lag phase (1998−2003, Fig. 8.3) and for a recent year (2013) during its linear

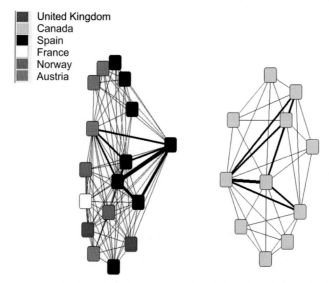

FIGURE 8.3 Co-authorship network for mountain cold-trapping during the lag phase (1998−2003). Each symbol represents an author with lines depicting ties between co-authors. Symbol color reflects the country with which an author was affiliated. The country key is placed in at the top left corner of the figure. Two clusters of co-authors are obvious: Canadian and European/United Kingdom clusters that focused their respective efforts on different mountain ranges. A core of consistently co-authoring researchers was obvious in both clusters (number of co-authored publications was indicated by the thickness of the ties). The European/United Kingdom cluster had a core of Spanish authors.

---

5. Along this line of thinking about scientific teams, Morris and Goldstein (2007) categorized publications produced by different types of scientific groups. At one extreme is a single authored paper (lone wolf) as might be written about a mathematical issue that required narrow expertise and minimal resources/equipment. Next on the continuum would be a publication from the typical small research team (wolf den) with little diversity of tasks among members. This team would typically be associated with a single laboratory or small group in an academic department. Next would be a product of a multiteam (wolf pack, perhaps 10−20 members) characterized by a distinct division of tasks among members as might be the case if several specialties were required to address the research question. At the highest level (big science or buffalo herd), many diverse teams might be engaged with a few core members coordinating the process. Unlike the implicit assumption of full intellectual ownership associated with authorship of a lone wolf or wolf den publication, all authors of a big science publication are not assumed to have full intellectual ownership of all aspects of the paper.

increasing phase (Fig. 7.4). The first thing to notice in both figures is that important attributes can be attached to nodes (authors). In the figures, authors from the same countries were identified by node color. During the initial lag phase for this innovation (Fig. 8.3), separate European and Canadian components focused on regional mountain ranges. By 2013, North Americans and Europeans were collaborating much more and had bridged to Asian researchers interested in contaminant movement in their own mountains and plateaus (Fig. 7.4). A pair of authors publishing on mountain cold-trapping in the Southern Alps of New Zealand formed a third component. Clearly, the network had evolved through time, integrating more authors into components and spreading geographically to encompass a larger number of high-altitude landscapes. Associating attributes with nodes (i.e., country in this example) facilitates understanding of the network structure and dynamics.

More than one attribute associated with authors can be used to gain insight about networks. The initial lag phase (1988−98) network for the Pollution-Induced Community Tolerance (PICT) innovation can be examined using author country and institution type (Fig. 8.4). Three components are

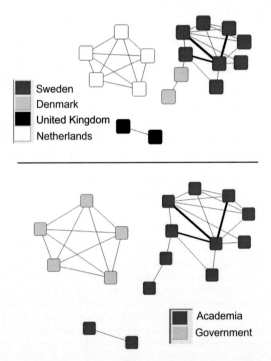

FIGURE 8.4 Co-authorship network for PICT during its lag phase (1988−98). Each symbol represents an author with lines depicting ties between co-authors. Tie thickness reflects the number of co-authored papers for the corresponding dyad. In the top and bottom panels, symbol color reflects the country and institution type with which an author was affiliated, respectively.

identifiable during the lag phase for this innovation. The largest has a core of Swedish academics who published several papers together. A pair of Danes, including a pendant author, also published with these Swedes. A set of Dutch government co-authors and pair of British academics also published on the PICT during its early stage of diffusion.

8.2.1.1   Whole Versus Ego Networks

If you glance back at the opening example about influences on 80 ecotoxicologist's *h*-indices, you will quickly see that it describes a network feature *from the vantage of a particular actor*. The ResearchGate site developers define the Reach metric using the second-person singular pronoun to make this clear. "RG Reach is a way to gauge the visibility of your work on ResearchGate. It reflects how many unique researchers can get notified when you add new research to your ResearchGate contributions page. Reach adds together your number of direct connections and the number of people connected to your work via your co-authors and project collaborators." RG Reach is a metric for what is called an ego network: it is not intended to be a metric that reflects some quality of a whole network.

Several ego network features can be illustrated using ResearchGate Reach and Fig. 8.2 (bottom, leftmost hexagonal network, (g)). Notice that actor C has six direct ties but actor A has only three (to B, C, and D). Actor A has three indirect ties (to E, G, and H) through actors B, C, and D. From the vantage of a particular actor, the Direct Reach index is a count of the number of direct ties, the Indirect Reach index is a count of the ties that your co-authors and collaborators have, and the RG Reach metric is the sum of the direct and indirect ties. The RG Reach metrics reflect direct ties and ties through close associates that might enhance the visibility of an individual's new work within the scientific community. Notice that Actors A and C in this example would have identical RG Reach metrics of six but the walks (i.e., paths through actors) for information movement would be different. The average number of actors that information would need to go through to get to A would be larger than for C. These types of differences in ego networks are important to understand.

Structural qualities of whole networks are the focus of many studies, and similar approaches exist for whole network analysis. For instance, it might be useful to quantify the average of the shortest paths between actors in the bottom two hexagonal networks shown in Fig. 8.2 ((g) and (h)). It is clear without resorting to any whole network metrics that they are quite different. Such differences influence the dynamics of the teams of individuals.

8.2.1.2   Centrality, Cohesion, and Subgroups

One important actor-associated quality of networks is centrality. Depending on the network being studied, an actor with high centrality might be more popular, possess higher prestige, carry more influence during group decision-making, or be more capable of exerting control during information flow than others with lower centrality (Borgatti et al., 2013). Another take on centrality is that it is "an index of the exposure of a[n actor] in the network ... the 'risk' of receiving whatever is flowing through the network" (Borgatti et al., 2013). As an example, an actor with high centrality might strongly influence how information (such as evidence supporting or questioning a research or risk hypothesis) diffuses through a network of colleagues. As a straightforward example, actor C of network (e) depicted leftmost in the middle row of Fig. 8.2 has much more influence than the other actors due to her central location. A less abstract example of contrasting centralities involves the individuals making up the Swedish/Danish academic component of the PICT network depicted in Fig. 8.4. Figure 8.5 is a closer look at this component with the total number of papers published by each author inserted inside each node symbol. The number of co-authored papers for author dyads is indicated by tie thickness and a number placed along the tie between multiple co-authoring scientists. The originator of the innovation (A1) produced the most publications. He published many with three co-authors (A2, A3, and A6) and together they formed the core of this network component. The A12 pendant had much lower centrality than author A1, and published only a single paper with co-author A9.

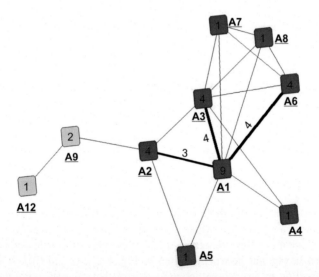

FIGURE 8.5 Swedish/Danish component of the PICT network shown in Fig. 8.4. The total number of papers published by each author is indicated within the node symbol and the number of co-authored papers for a dyad of authors indicated by line thickness and a number placed along the tie. Each author is identified with an underlined letter A followed by a unique number. For example, the innovator (A1) published nine papers on the subject. Three, four, and four were published with the indicated Swedish co-authors. In contrast, author A12 published only one paper with a fellow Dane (A9), and was situated less centrally than author A1 in the network component.

Centrality can be measured in several ways, one of which has already been touched upon. The simplest is to quantify degree, that is, the number of direct ties to a node. In this example, the respective degree centralities of A1 and A12 are 7 and 1, respectively. Other degree-based measures are available such as the RG Reach described in the opening example and Section 8.2.1.1. That centrality metric takes into consideration the number of actors to whom an ego was tied plus the number of actors to whom those actors were tied. This is a two-step reach metric which is the sum of all actors within two tie steps of an ego. It can be expressed as a simple count or as a percentage of all actors in the network who are within two steps of the ego. To adjust this percentage for matrix size, it is often calculated as $100 \times$ [the number of actors within 2 steps/(total number of actors in the network $-$ 1)]. Another common measure of actor centrality is betweenness centrality, that is, how often a particular actor is along the shortest path connecting two other actors (Borgatti et al, 2013).[6] The betweeness centrality metric reflects an actor's capacity to serve as a connector. Newman (2001) provided a helpful vantage for understanding betweenness that emphasizes the role of an ego in network functioning, "The [actors] with the highest betweenness also result in the largest increase in typical distance between others when they are removed." The betweennesses of A9 and A12 are 8 and 0, respectively. Although distant from many actors, A9 had good potential to act as a connector as indicated by its relatively high betweenness: he connected A12 to all other actors. Centrality measures for each actor (Table 8.1) indicate that A1 was most influential in the network, having the highest potential for being a connector.

Other centrality measures for undirected networks include closeness, eigenvector, beta, and k-step centralities. Centrality metrics normalized for the size of networks are routinely applied in cases where the networks being compared differ greatly in size.

Centrality matters in co-authorship networks such as the steel engineering literature studied by Uddin, Hossain, and Rasmussen (2013). Uddin et al. found that the citation counts for three-authored papers were positively influenced by the degree and betweenness centralities of the co-authors in their respective authorship networks. In a scientific co-authorship study by Abbasi, Hossain, and Leydesdorff (2012), betweenness centrality of an established scientist was related positively with a new scientist's preference to associate with them. As a final example, results of the opening survey (Fig. 8.1) suggested

---

6. The shortest path between a pair of actors is also called the geodesic distance. Estimating geodesic distances can be problematic for networks with more than one component because unconnected actors have undefined (or defined as infinite) distances. Borgatti et al. (2013) indicate that reciprocals of geodesic distances can be used to produce indices in such instances because the reciprocal transformation produces values of 0 between actors in separate components.

TABLE 8.1 Centrality Metrics for the Individual Actors Depicted in Fig. 8.5 as Calculated With the UCINET V6.577 Network Analysis Program (Borgatti et al., 2002)

Actor	Degree	Betweenness
A1	7	12.5
A2	4	14.5
A3	3	7.5
A4	2	0.0
A5	2	0.0
A6	4	0.0
A7	4	0.0
A8	4	0.0
A9	2	8.0
A12	1	0.0

that the combined productivity and citation counts of individual ecotoxicologists were positively related to degree centrality expressed as Direct Reach.

Ego centrality can be used to further understand the linear trend in lag times for ecotoxicology innovations (Fig. 7.5). One possible contributor to the trend might be that innovator centrality in the initial lag phase co-author network influences how quickly or slowly an innovation reaches the tipping point after which the annual rate of new author involvement increases sharply. The field of pollution science lacked an established social organization initially, and an innovation required considerable time to gradually emerge. Early publications on the topic were written by authors affiliated with other disciplines. Opportunities for establishing ties to others were fewer in the 1960s compared to later years when rapid advances in information technology occurred. To explore the possibility of innovator centrality being related to the lag time trend, the two-step reach was estimated for each of the 18 innovators and plotted against the year that the innovation was introduced into the literature. Supporting the explanatory hypothesis posed at the beginning of this paragraph, innovator centrality did increase steadily from the 1960s to the present (Fig. 8.6). Several innovators had very low centralities, especially during the first three decades. Three early innovators published alone and only once on the topic during the lag period. Speculating, this analysis of two-step reach supports the hypothesis that the general increase in innovator centrality through time corresponded with the shortening of the lag period through time.

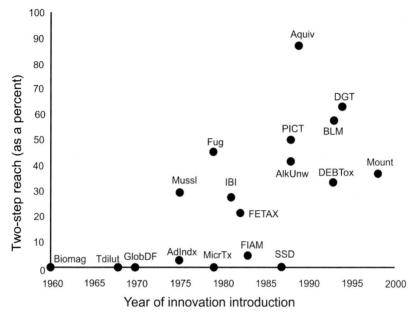

FIGURE 8.6   The shift in innovator two-step reach (expressed as a percentage) over time. This ego centrality measure increased as the calendar year of innovation introduction increased (Pearson $\rho = 0.591$; 95% highest density credible interval $= 0.269 - 0.880$; probability that $\rho > 0$ is 0.996). Innovation abbreviations are the same as described in the legend of Fig. 7.5.

Another important network quality is cohesion, the extent to which actors in a network are tightly connected. Density, a common cohesion metric, is the total number of ties in a network divided by the total number of possible ties for a network of its size. (The total possible number of dyadic ties for an undirected network is $n (n - 1)/2$, where n is the number of actors in the network.) A high-density network is more tightly connected than a low-density network. Because very large differences in network size influence the calculated values of the density metric, other measures can be used instead if cohesion is being compared for networks of very different sizes. Often, the average of the degrees computed for all of the individual actors in the network is used to quantify cohesion. In the example depicted in Fig. 8.5, the average degree estimate of cohesion for the whole network is 2.118.

An example of network cohesion importance involves the evolution of scientific co-authorship networks (Barabási et al., 2002). Through time, the average degree for a co-authorship networks increased and node separation decreased: co-authorship networks become more tightly knit as they mature through time.

Other attributes involving cohesion are the number of components in a network and how regions of a network might differ relative to density. All else being equal, a network with few components is more cohesive than one with many components. And two networks with the same number of actors and components might have similar connectedness but, if one has a very large central component, it will have higher connectedness than the network with actors more evenly distributed among components.

A clustering coefficient might also be calculated by a researcher to quantify unevenness in network densities of actors. A cluster coefficient for a scientific co-authorship network is "the probability that two scientist's coauthors have themselves coauthored a paper ... in topological terms, it is a measure of the density of triangles in a network, a triangle being formed every time two of one's collaborators collaborate with each other" (Newman, 2004). In its rudimentary form, it is the number of components divided by the number of actors in the network, c/n. The PICT co-authorship network in Fig. 8.4 has an estimated component ratio of 3 components/17 actors or 0.176. This ratio is often normalized using $(c - 1)/(n - 1)$ if networks being compared vary substantially in size (Borgatti et al., 2013). The normalized component ratio for the PICT network (Fig. 8.4) would be 2/16 or 0.125. When comparing networks, a lower component ratio reflects higher cohesion.

Analysis of cohesiveness of the initial lag phase networks of the ecotoxicology innovations shown in Fig. 7.5 provides further insight about network changes co-occurring with the progressive shortening of the lag period from 1960 to the present. Whole network cohesion (normalized component ratio)

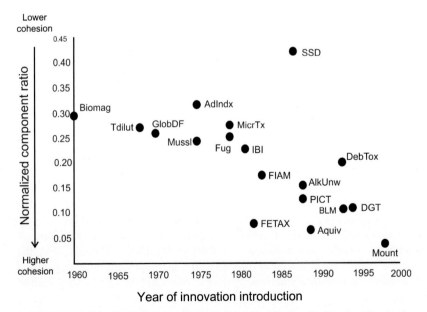

FIGURE 8.7 The increase in network cohesion with increase in the calendar year of innovation introduction. Innovation abbreviations are the same as described in the legend of Fig. 7.5.

for the 18 innovations was plotted against the year of introduction (Fig. 8.7), producing a clear trend of increasing network cohesion with time. The species sensitivity distribution (SSD) innovation appears as an outlier. Speculating, the SSD approach is very easily implemented using existing electronic databases, is a recent modification of an established US EPA method for setting water quality criteria, and is currently recommended in numerous ecological risk assessment guidance documents. The net effect of these qualities is that it takes minimal interaction among individuals to implement the SSD, and unconnected groups of individuals frequently apply it during diverse risk assessment activities.

Overall, the shortening of the innovation diffusion lag periods was accompanied by an increase in whole network cohesion and innovator centrality. It is plausible, but unproven, that the increased innovator centralities and whole network cohesions resulted from the general trend toward more collaborative science, maturing of ecotoxicology as a science, advances in information technology, or a combination of the three. Likely, these network qualities reflect temporal trends and are correlated to the shortening of the diffusion lag period though time. Speculating again, the shortening of the lag phase seems to indicate a transition through time from internal to an external influence model dynamics.

Still other cohesive substructures such as cliques or factions can exist within a network and methods exist to characterize them. The meaning of clique is more specific in social network analysis than in the general vernacular. A clique as "a subset of actors in which every actor [has a tie with] every other actor in the subset and it is impossible to add any more actors to the clique without violating this condition" (Borgatti et al., 2013). Knoke and Yang (2008) define a clique more formally as being "... the maximal complete subgraph of three or more [actors], all of which are directly connected to one another, with no other [actor] in the network having direct ties to every member of the clique. Hence, the geodesic distance for every pair of nodes in a clique is 1." The Dutch government co-authors in Fig. 8.4 are a clique because they all have geodesic distances of 1 and no other actor in the network has direct ties with all of them. The component of 10 Swedish/Danish academics also contains a clique; however, some co-authors in the component are not in that clique. Co-authors designated A1, A3, A6, A7, and A8 in Fig. 8.5 form a clique because they all form dyads with each other with geodesic distances of 1. In contrast, the Danish co-authors (A9 and A12) are clearly not in this clique because their geodesic distances from several members of the clique are greater than 1. Also, co-authors A2, A4, and A5 have geodesic distances greater than 1 with some clique members so they also are not members.

A faction is slightly different from a clique in social network analysis. With cliques, an individual could be a member of more than one clique as long as the conditions for membership were met for each clique. Factions are found in networks using the rule that an actor can belong to only one faction (Borgatti et al., 2013). To identify factions, an *a priori* number of factions is defined and then various techniques based on densities applied to best divide the network into that number of factions.

Various matrix-based methods exist for identifying clusters, cliques, factions, and other substructures. However, they require more explanation than possible in this short chapter. The interested reader is referred to Borgatti et al. (2013) or Knoke and Yang (2008) for more computational details.

8.2.2 Actor Roles and Influences in Networks

"Territorial science...blocks the flow of metaphors and the development of new theories. Distrust and disinterest in anything outside one's subdiscipline supports surrogates for theory."

Gigerenzer (2000)

The network methods described above were developed to better understand the nature of social systems and the roles played by associated actors. Other relevant topics explored in other chapters can also be incorporated into network analysis to add further understanding. For instance, homophily (see Section 6.1.2) is often analyzed in social networks to determine why some actors choose to associate with others. The qualities of small groups that lead to suboptimal decision-making (see Sections 6.3 and 6.4) is a second instance in which network analysis can provide insight about how information exchange can go awry. The simple comparison of seven-member networks shown in Fig. 8.2 suggests network qualities that might lead to groupthink. Other instances amenable to network analysis are the identification of opinion leaders in attempts to facilitate innovation diffusion (see Section 7.3.1), more fully understanding the strength of weak ties (see Section 7.3.3), and the existence of critical tipping points in adoption curves (see Fig. 7.3).

Important roles such as leader, follower, or bridge can be identified and understood within networks. Degree, betweenness, and density can help identify informal opinion leaders or quantify effectiveness of an official leader. Network analysis can also identify those who are most capable of obstructing innovation or evidence flow. Analyses identifying clique membership might allow selection of the best advisory committee composition by ensuring that a clique or faction is not unintentionally overrepresented.[7] Also ego network analysis, such as that facilitated by ResearchGate, can identify patterns of

---

7. More than three decades of engagement on advisory committees suggests to me that clique or faction overrepresentation is one of the most common impediments to optimal advice giving and decision-making. Those charged with creating committees often try to get the best individuals but give short shrift to creating the best mix of individuals. Bias toward faction overrepresentation is worsened by the well-intended tendency to select members of past teams that have a record of cooperating to achieve results like those desired from the proposed committee. Often such a team is a clique or faction, and only one individual from it is actually needed to represent its views and evidence.

contribution or foster effectiveness. Creating or protecting bridges that facilitate information diffusion between and within networks is another possibility. For example, the bridge shown in Fig. 7.4 might be provided incentive by funding agencies to continue to connect disparate scientists or publically praised at his home institution for being an exemplary scientist.

In Chapter 6, Social Processing of Evidence: Commonplace Dynamics and Foibles (Section 6.1.2), we examined how individuals tend to obey the homophily principle of "similarity breeds connection" (McPherson, Smith-Lovin, & Cook, 2001). Individuals tend to associate with others similar to themselves and are less prone to associate with those who are dissimilar. This principle can be detected and quantified in networks.

> ... *if two people have characteristics that match in a proportion greater than expected in the population from which they are drawn or the network of which they are part, then they are more likely to be connected ... [this] homophily principle ... applies equally to groups, organization, countries, and other social units.*
>
> Kadushin (2012)

Kadushin (2012) defines several general processes fostering network homophily: "... (1) the same kinds of people come together; (2) people influence one another and in the process become alike; (3) people can end up in the same place; (4) and once they are in the same place, the very place influences them to become alike."

Homophily influences network self-shaping, collaboration, information flow, trust, and engagement. As an example, Hâncean and Perc (2016) found pervasive homophily in the sociology literature of east European countries (Poland, Romania, and Slovenia). In another study of homophily in scientific networks, Kretschmer, Beaver, Ozel, and Kretschmer (2015) applied network methods to highlight the role of homophily in the self-shaping and functioning of male and female co-authorship collaborations.

As a word of caution before proceeding further, what appears to be homophily-based network structuring might be due to the Matthew (accumulated advantage) effect (Merton, 1968). Matthew here refers to a parable in the Gospel According to Matthew (Matthew 25:29). "For everyone who has will be given more, and he will have an abundance. But the one who does not have, even what he has will be taken away from him." According to the Matthew effect, how much individuals or groups of scientific collaborators are rewarded is based on what they have already accomplished. Past success warrants further rewards, and the lack of success warrants fewer or no rewards. Given this effect, heterogeneity will emerge in some networks as individuals find it to their advantage to remain affiliated with successful individuals or groups. A scientist with modest accomplishments might strive to affiliate with a very prominent individual in the field with the hopes of garnering more status for themselves. The Matthew effect can produce substructures in collaboration networks that can appear superficially to result from homophily.

Measures of homophily include Yule's Q, E-I (Group External−Group Internal Ties) index, and conventional correlation coefficients. Yule's index will be illustrated first because it is quite straightforward and, unlike the E-I index, is insensitive to differences in group sizes. Yule's Q is calculated from a tabulation of an actors' ties with others who are the same (a) or different (b), and the number of "no ties" of actors with others who are the same (c) or different (d) (Borgatti et al., 2013).

$$Yule \; Q = \frac{ad - bc}{ad + bc}$$

A Q value of 0 reflects no apparent relationship between actor attribute similarity and tie status. Perfect heterophily or homophily would produce Yule's Q values of -1 and $+1$, respectively.

The bottom panel of Fig. 8.4 shows a small network of authors who were employed at academic or government institutions. Notice that no member of one type published with a member of the other type, so homophily was high as reflected in a corresponding Yule's Q value of 1. Figure 8.8 of the lag phase mountain cold-trapping co-authorship network is similar to the network shown in Fig. 8.3 but the actor roles are now identified by the type of institute with which they were associated. In contrast to Fig. 8.4 (bottom panel), the government authors are scattered among the academic authors in this network. This weak homophily pattern is reflected by a low Yule Q value of 0.1316. Relative to author's institute type, there is much more

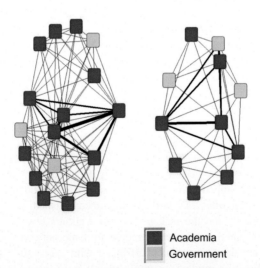

Academia
Government

FIGURE 8.8 Co-authorship network for mountain cold-trapping during the lag phase (1998−2003). Each symbol represents an author with lines being ties connecting co-authors. Symbol color reflects the role of each co-author, i.e., academic (purple) or employed by a government agency (blue).

homophily in the PICT network than in that of mountain cold-trapping. Examples intermediate between the PICT and mountain cold-trapping networks were those of the additive index (Yule $Q = 0.7882$) and alkaline unwinding assay (Yule $Q = 0.3898$) during their lag phases.

The E-I index was developed by Krackhardt and Stern (1988) to study the role of friendship networks during organizational crises. Friendship ties between actors were classified as being inside (internal) or outside (external) of an organizational subunit. The index was simply,

$$\text{E} - I \text{ Index} = \frac{E - I}{E + I}$$

where E = the number of external friendship ties and I = the number of internal friendship ties. Notice that E and I correspond to b and a, respectively, in the above equation for Yule's Q. So, Yule's Q includes numbers with no ties but the E-I index does not. E-I index scores can range from -1 (all ties internal) to $+1$ (all ties external). Notice that, opposite of the Yule Q, negative values for the E-I index indicate more internal ties (i.e., homophily). Because the numbers with no ties are ignored (c and d in Yule's Q), the minimum and maximum E-I index scores can vary depending on the relative sizes of the groups being examined. A rescaling of E-I index scores might be needed to compare networks as done, for example, by the UCINET network analysis program (Borgatti, Everett, & Freeman, 2002).

The question might be asked about the changes in homophily through time for the 18 ecotoxicology innovations. Perhaps, homophily decreases through time relative to the professional roles or country affiliations of co-authors. To address this hypothesis, an actor's role and country were defined by their institutional affiliation as reported on the title page of each publication. Professional role were academic, government, consulting, non-government organization (NGO), or industry. In most cases, government agencies were environmental regulatory agencies and industries were those subject to environmental regulation. Rescaled E-I index values for an actor's professional role (mean = -0.477; 95% highest density credible interval = -0.607 to -0.344) and country affiliation (mean = -0.770; 95% highest density credible interval = -0.890 to -0.651) indicated pervasive homophily. Homophily relative to country was more pronounced than that of professional role. However, inconsistent with the above hypothesis, plots of rescaled E-I index scores versus calendar year provided no evidence for a temporal trend in homophily based on professional or country affiliation.

8.3   NETWORKS OF SCIENTISTS

... article authors ... represent the human creators of research. As such, their behavior and social organization drive all other aspects of the specialty.

Morris and Goldstein (2007)

The co-authorship network of scientists represents a prototype of complex evolving networks.

Barabási et al. (2002)

Many sorts of networks exist with citation and organizational networks most likely being the most familiar to ecotoxicologists and ecological risk assessors. Co-authorship networks are emphasized here because they reflect stronger bonds among members than do citation networks (Uddin et al., 2013) and also organizational networks vary so widely among institutes and agencies. The existence of an enormous bibliographic body of information about scientific paper authoring also makes co-authorship networks attractive tools for understanding collaboration in networks of scientists.

Because of this focus on simple co-authorship networks, a bit more detail is needed before proceeding. Foremost, not all individuals publishing on a topic play the same role although this might not be obvious in the analyses here. Price and Gürsey (1975) and later Braun, Glänzel, and Shubert (2001) apply a straightforward scheme to categorize authors in a scientific discipline as being newcomers, transients, terminators, or continuants. During any particular publication interval, there will be newcomers who appear for the first time in the literature. A student publishing their Master's thesis research will likely be a newcomer. Some will be transients who publish just that once and never again. This might be the case for an individual whose primary focus is another science, but nonetheless, they offer a novel idea about a topic of interest in another field. For example, a mathematician might co-author with an ecologist about a quantitative solution to a long-standing question in landscape ecology. There will also be terminators who published previously but never again after that time period. The final category includes continuants who published before and will continue to publish after that period. A successful academician who has built a career studying a particular issue in ecotoxicology would likely be a continuant. She might have a publication record including student co-authors (newcomers) who might go on to become terminators, transients, or hopefully, continuants. The combined activities of these four types of contributors give form to the literature of a science.

The evolution of co-authorship networks through time usually results in a core-periphery structure. Highly productive actors become more influential with time (Crane, 1969). In the time frame of decades, members of the core and periphery may move out of one and into the other, and new members might arrive and old members leave. Despite these dynamics, a core-periphery structure tends to be retained (Cugmas, Ferligo, & Kronegger, 2016) with a core of multiple co-authoring continuants surrounded by a swarm of less prominent authors. Although they are not fully developed, the lag phase networks shown in Figs. 8.3 and 8.4 have emerging core-periphery structures. A better illustration of core-periphery structure is that for the

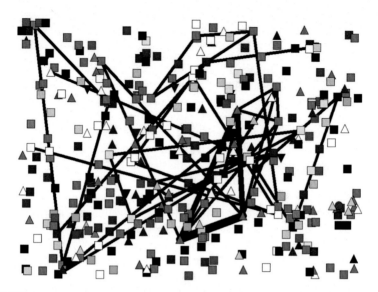

FIGURE 8.9 Co-authorship network for biomagnification during the most recent year analyzed for the linear phase (2013). Single co-authorship ties are not shown here, so that it is easier to see the core of multiple co-authors. (There was only one isolate in the network with no ties to another author.) Node color and shape show the 31 countries of affiliation and six roles of the network authors. China (dark green) was the most prominent country in the core, and academia (square) and government (triangle) were the dominant roles of core actors.

biomagnification concept during the most recent part of its linear phase (Fig. 8.9). By that time, the concept had been integrated into the ecotoxicology literature for more than half a century. To see through the clutter in this network diagram, connecting lines were omitted for co-author dyads with only one publication together. Lines drawn between author dyads with multiple co-authored papers had line thicknesses that reflected how many publications the dyad co-authored together. This biomagnification network has a clear core surrounded by a swarm of peripheral authors.

Relative to knowledge flow and consensus building, such self-organizing core-periphery structures are believed to be superior to tightly knit networks of scientists dominated by one or a few cliques or factions (Johnson et al., 2003). Informal communication is fostered about "findings, research-in-progress, and research techniques" (Crane, 1969) in core-periphery network structures. This structure reduces the risk of negative stereotyping of out-groups or marginalization of individuals as might emerge with ingroup groupthink. Although actors in the core are more connected to each other than to those in the periphery, the dynamic nature of invisible college structure provides a healthy means by which motivated actors can increase their contributions to the science by engaging with successful core actors.

8.4 CONCLUSIONS

8.4.1 General

To introduce the reader to the importance of social network dynamics in eco-toxicology, the opening example demonstrated that an ecotoxicologist's direct reach influences their impact as reflected in the h-index. Beyond the influence of networks on individual careers, scientist networking benefits society by improving knowledge transfer among scientists, broadening the expertise that can be focused on a question, and increasing the chance that knowledge will be available to address transnational problems and problems important in less developed countries.

Ties connecting actors in a network can reflect relational events or roles, or cognitive states. They can be analyzed from the vantage of an ego or the whole network, and metrics exist for both vantages. Important attributes are centrality, cohesion, and substructuring. As an ecotoxicology example, innovator centrality increased with time in our analysis of innovation lag phase duration, suggesting an ego network quality correlated with the shortening in lag time from the 1960s to the present (Fig. 8.6). Changes in whole network cohesion for the 18 ecotoxicology innovations also were correlated with the shortening of innovation lag time duration (Fig. 8.7).

Network components, cliques, and factions influence how knowledge moves among or decisions are made by collections of individuals. Attributes can be assigned to actors in the network such as gender, age, nationality, primary language, scientific discipline, or type of institute providing employment. The homophily principle and Matthew effect can also manifest in networks of scientists, influencing knowledge transfer and decision-making. Although they could plausibly be related to the changes in lag time duration for the 18 ecotoxicology innovations, no evidence was found for any relationship between co-author professional role or country homophily and lag time for an innovation adoption curve.

A core-periphery structure is common for members of a scientific discipline. In contrast to networks with strong cliques or factions, core-periphery networks are believed to allow more knowledge flow, more efficient consensus building, and more opportunity for an actor in the periphery to transition into the core.

8.4.2 Emerging Possibilities of Crowdsourcing

An emerging field of evidence processing by very large groups must be mentioned before ending this chapter. The reader might recollect that Francis Galton was mentioned in Chapter 2, Human Reasoning: Everyday Heuristics and Foibles, relative to the correct interpretation of regression results. He is less well known for an observational experiment that he conducted at the 1884 West of England Fat Stock and Poultry Exhibition (Galton, 1907). Entertaining lotteries were common at such rural gatherings and, at this particular exhibition, a prize was to be given to the individual who best guessed the weight of dressed meat produced from a displayed oxen. Galton gathered 787 legible lottery tickets with individual guesses— from experts and novices alike—and calculated a collective average of 1197 pounds. This crowd-generated average was only 1 pound below the actual figure. He concluded that "This result is, I think, more credible to the trustworthiness of a democratic judgment than might have been expected." Surowiecki (2005) and others refer to this approach of generating sound judgments, estimates, or collaborative products from large groups as crowdsourcing. With the advent of powerful IT capabilities, crowdsourcing is emerging as a new means of collaboration (e.g., Michelucci & Dickinson, 2016; Surowiecki, 2005) and a variety of tools have emerged to facilitate it (e.g., Paolacci, Chandler, & Ipeirotis, 2010; Turban, Liang, & Wu, 2011)

> *Collective decisions are most likely to be good ones when they're made by people with diverse opinions reaching independent conclusions, relying primarily on their private information.*
>
> Surowiecki (2005)

Diversity, independence, and decentralization are essential to getting good results from crowdsourcing (Surowiecki, 2005). The value of a diversity of backgrounds and experiences should be no surprise given our previous discussions of the best group decision-making practices. Diversity provides a wide range of experiences and broad knowledge base from which to draw. The same is true for independence of opinions: initial independence of opinions was a crucial feature of the Delphi expert elicitation methods described in Chapter 6, Social Processing of Evidence: Commonplace Dynamics and Foibles (Section 6.5). Recollect from Chapter 2, Human Reasoning: Everyday Heuristics and Foibles (discussion of informational mimicry) also that there is a risk of information cascades if independence is not maintained and social learning dominates. Decentralization assures that a wide network of contributors is involved: dominance of one or a few subgroups would diminish the independence of individual contributions. One instance in which the benefit of decentralization has occurred is Wikipedia. Many individuals contribute to entries, bringing together a body of information on a subject likely unachievable by the ordinary individual. On average, a contribution would be diminished if only one or a few subgroups in a field decided what material to include in an entry. Indeed, pages have been closed when input appeared one-sided.

REFERENCES

Abbasi, A., Hossain, L., & Leydesdorff. (2012). Betweenness centrality as a driver of preferential attachment in the evolution of research collaboration networks. *Journal of Informatics, 6*, 403−412.

Barabási, A. L., Jeong, H., Néda, Z., Ravasz, E. M., Schubert, A., & Vicsek, T. (2002). Evolution of the social network of scientific collaborations. *Physica A, 311*, 590−614.

Boccaletti, S., Latora, V., Moreno, Y., Chavez, M., & Hwang, D.-U. (2006). Complex networks: Structure and dynamics. *Physics Reports, 424*, 175−308.

Borgatti, S. P., Everett, M. G., & Johnson, J. C. (2013). *Analyzing social networks.* Los Angeles: SAGE Publications.

Borgatti, S. P., Everett, M. G., & Freeman, L. C. (2002). *Ucinet 6 for Windows: Software for social network analysis.* Harvard: Analytic Technologies.

Braun, T., Glänzel, W., & Shubert, A. (2001). Publication and cooperation patterns of the authors of neuroscience journals. *Scientometrics, 51*, 499−510.

Crane, D. (1969). Social structure in a group of scientists: A test of the "invisible college" hypothesis. *American Sociological Review, 34*, 335−352.

Cugmas, M., Ferligo, A., & Kronegger, L. (2016). The stability of co-authorship structures. *Scientometrics, 106*, 163−186.

Galton, F. (1907). Vox populi. *Nature, 75*, 450−451.

Gigerenzer, G. (2000). *Adaptive thinking. Rationality in the real world.* Oxford: Oxford University Press.

Hâncean, M.-G., & Perc, M. (2016). Homophily in coauthorship networks of East European sociologists. *Nature Scientific Reports, 6*, 36152.

Hirsch, J. E. (2007). Does the h index have predictive power? *Proceedings of the National Academy of Sciences of the USA, 104*, 19193−19198.

Jha, Y., & Welch, E. W. (2010). Relational mechanisms governing multifaceted collaborative behavior of academic scientists in six fields of science and engineering. *Research Policy, 39*, 1174−1184.

Johnson, J. C., Boster, J. S., & Palinkas, L. A. (2003). Social roles and the evolution of networks in extreme and isolated environments. *Journal of Mathematical Sociology, 27*, 89−121.

Kadushin, C. (2012). *Understanding social networks. Theories, concepts, and findings.* Oxford: Oxford University Press.

Katz, J. S. (1994). Geographical proximity and scientific collaboration. *Scientometrics, 31*, 31−43.

Krackhardt, D., & Stern, R. N. (1988). Informal networks and organizational crises: An experimental simulation. *Social Psychology Quarterly, 51*, 123−140.

Kretschmer, H., Beaver, D. deB., Ozel, B., & Kretschmer, T. (2015). Who is collaborating with whom? Part II. Application of the methods to male and female networks. *Journal of Informatics, 9*, 373−384.

Knoke, D., & Yang, S. (2008). *Social network analysis (2nd ed.).* Los Angeles: SAGE Publications.

McPherson, M., Smith-Lovin, L., & Cook, J. M. (2001). Birds of a feather: Homophily in social networks. *Annual Review of Sociology, 27*, 415−444.

Merton, R. K. (1968). The Matthew effect in science. *Science, 159*, 56−63.

Michelucci, P., & Dickinson, J. L. (2016). The power of crowds. *Science, 351*, 32−33.

Morris, S. A., & Goldstein, M. L. (2007). Manifestation of research teams in journal literature: A growth model of papers, authors, collaboration, coauthorship, weak ties, and Lotka's law. *Journal of the American Society for Information Science and Technology, 58*, 1764–1782.

Newman, M. E. J. (2001). Scientific collaboration networks. II. Shortest paths, weighted networks, and centrality. *Physical Review E, Statistical, Nonlinear, and Soft Matter Physics, 64*, 016132.

Newman, M. E. J. (2004). Coauthorship networks and patterns of scientific collaboration. *Proceedings of the National Academy of Sciences of the USA, 101*(Suppl. 1), 5200–5205.

Paolacci, G., Chandler, J., & Ipeirotis, P. G. (2010). Running experiments on Amazon Mechanical Turk. *Judgment and Decision Making, 5*, 411–419.

Price, D. D., & Gürsey, S. (1975). Studies in scientometrics. I. Transience and continuance in scientific authorship. *Ciência da Informação, 4*, 27–40.

Rodriguez, M. A., & Pepe, A. (2008). On the relationship between the structural and socioacademic communities of a coauthorship network. *Journal of Informetrics, 2*, 195–201.

Surowiecki, J. (2005). *The wisdom of crowds.* New York: Anchor Books/Random House.

Turban, E., Liang, T.- P., & Wu, S. P. J. (2011). A framework for adopting collaboration 2.0 tools for virtual group decision making. *Group Decision and Negotiation, 20*, 137–154.

Uddin, S., Hossain, L., & Rasmussen, K. (2013). Network effects on scientific collaborations. *PLoS One, 8*, e57546.

Section 4. Conclusion

Chapter 9

Conclusion

9.1 THE ELEPHANT-IN-THE-ROOM

The book's preface opens with Krylov's elephant-in-the-room fable because I felt at midcareer like the museum goer who had overlooked a major item in a gallery filled with absorbing specimens. For decades, I published on technical and scientific aspects of the new science of ecotoxicology. These publications included introductory and advanced textbooks that framed ecotoxicological findings around scientific principles but eschewed "off topic" issues that seemed to strongly influence how ecotoxicology knowledge was created and applied. Occasionally while reading manuscript reviews, phrasings would appear that suggested the influence of factors other than scientific soundness on final recommendations. Perhaps, a comment would be made about an important citation that was overlooked, someone's research findings that were not discussed, or conclusions being inconsistent with the current stance of a regulatory agency. I attended scientific meetings and sat on numerous panels addressing environmental pollution and risk issues. As committees lumbered toward completion of their assigned tasks, their collective reasoning was often swayed by indistinct but powerful influences unrelated to scientific or logical soundness. For the sake of civility, an excellent point might have been abandoned when it was not directly in line with the conclusion preferred by a few dominant members of a panel. Frequently, informal discussions with individual members after completion of a committee's task would reveal that a unanimous consensus was only superficially so. As student presentations were judged at meetings, something not on the judging forms would occasionally have a powerful influence on the outcome. And awards came often to some deserving researchers while other outstanding individuals remained unacknowledged. Unquestionably, vague influences on scientific and regulatory decision-making remained inadequately defined.

Approximately 20 years ago, the discomfort associated with my—and apparently other's—lack of understanding of these unacknowledged, but powerful processes grew acute enough that I began to gradually disengage from many expected midcareer activities and began a deliberate, private inquiry from a different vantage. While engaged at meetings, I paid increasing attention to the cognitive psychology and sociology of activities. I found myself

The Nature and Use of Ecotoxicological Evidence. DOI: https://doi.org/10.1016/B978-0-12-809642-0.00009-1
247

doing the same when reviewing publications and regulatory documents, or participating in advisory panels. In terms of Krylov's elephant-in-the-room fable, I tried to broaden my perspective by stepping out of the natural history gallery to browse the anthropology, sociology, and psychology wings of the museum. Returning to the natural history gallery with a fresh eye, I could see the general shape of the elephant in the room for the first time. The elephant's chief features were described in this short book and summarized below. A more refined treatment is left to others with more sophisticated psychology and sociology training.

The museum goer in Krylov's fable was asked several questions by his friend: "...But did you see the elephant? What did you think it looked like? I'll be bound you felt as if you were looking at a mountain." For myself, I can say that I now see the elephant, and indeed, it is huge (Fig. 9.1). And its shadow moves every day across all other items in the ecotoxicology gallery. That I previously had not seen it more clearly was to my intellectual disadvantage. It would have saved much effort in deciding which activity to spent time on and planning how to most effectively work through the activities I chose. This book is my attempt to describe the elephant-in-the-room to others, and in doing so, perhaps spare them some wasted effort associated

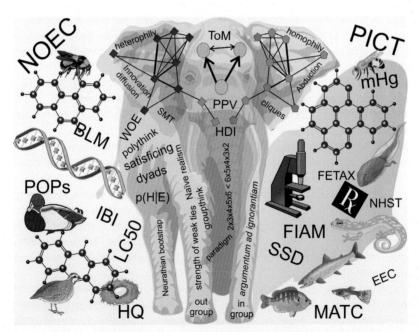

FIGURE 9.1 The elephant in the ecotoxicology gallery. BLM, biotic ligand model; EEC, estimated environmental concentration; FETAX, frog embryo teratogenesis assay - Xenopus; FIAM, free ion activity model; HQ, hazard quotient; IBI, Index of Biotic Integrity; NOEC, no and observed effect concentration; POPs, persistent organic pollutants; SMT, self-managing team; SSD, species sensitivity distribution; WOE, weight-of-evidence.

with not understanding its importance. My impression as I finish my career is that the elephant still looms prominently in a gallery filled with remarkable ecotoxicology facts and phenomena. Hopefully, this short book has convinced a few readers that it is having a major influence on how 21st century pollution issues are being addressed. And as such, it deserves careful study and consideration.

9.2   THE NATURE OF ECOTOXICOLOGY EVIDENCE

...though I cannot tell why this was exactly; yet, now that I recall all the circumstances, I think I can see a little into the springs and motives which being cunningly presented to me under various disguises, induced me to set bout performing the part I did, besides cajoling me into the delusion that it was a choice resulting from my own unbiased freewill and discriminating judgement.

Melville (1851)

A central theme in this short book was identifying tendencies and predispositions that delude us when making judgments. Perhaps, as expressed in the quote above, the exploration of these foibles in their "various disguises" instilled in the reader a better awareness of the associated problems and suggested ways of reducing their influences on our "unbiased freewill and discriminating judgment."

The opening chapter chronicled the emergence of pollution as a critical event in history, identifying the 20th century as the turning point for humankind. Human population size, densities, and distribution emerged as major contributors to widespread environmental degradation. Our increasingly effective abilities to extract resources and energy from the environment accelerated this degradation. By the close of the 20th century, knowledge of human impacts on the environment and ways of attenuating them became crucial for our continued well-being. Now, social and economic progress during the Anthropocene epoch requires evidence to be generated, communicated, and applied optimally. To accomplish this, more concerted and conscientious effort is needed to avoid inherent errors in individual and collective reasoning during evidence interpretation.

Commonplace heuristics create obstacles to optimally generating and organizing ecotoxicological knowledge. Useful in many everyday instances, they compromise the soundness of scientific and technical inferences if left unchecked. Given the complexity of modern ecotoxicology, it is now essential to understand the limits imposed by these heuristics and to actively minimize their influence. Unfortunately, many ecotoxicologists who automatically rely on them in their daily decision-making remain unaware that they are misapplying them in their professional judgments. As inopportune a truth as it might be, "common sense" and "global introspection" are insufficient in many aspects of current scientific reasoning and risk assessment.

An active understanding of the relevant heuristics outlined in Chapter 2, Human Reasoning: Everyday Heuristics and Foibles, can help lower the risk of misjudging from ecotoxicological evidence. Errors associated with estimating probabilities are especially difficult to avoid, and rote application of conventional statistical tests does not remedy this problem. More attention to and application of basic probability calculus is required to avoid mistakes. Reframing questions in terms of natural frequencies often facilitates sound inferences about the plausibility of alternative explanations or the risks of adverse impacts.

Numerous errors in inference emerge during unthoughtful application of popular statistical methods. A central example is the pervasive misinterpretation of P-values. Minor changes in how frequentist statistical analyses are conducted could improve the soundness of inferences. Examples include more consideration of Type II error and effect size, the estimation of positive and negative predictive value, framing of null hypotheses in more meaningful terms than the nil hypothesis, and wider application of confidence intervals where appropriate. Shortcomings of the frequentist approach have opened up the field to potentially useful Bayesian and information-theoretic methods.

Legitimate rules for conducting science vary widely despite the popular misconception that there is but one valid scientific method. That having been said some methods are more effective than others at reducing the influence of cognitive errors. Passé approaches such as precipitate explanation, enhanced belief based on repeated weak testing, and misinterpretation of falsification exercises still appear sporadically and muddle scientific progress in ecotoxicology. Deviation from sound scientific practice impedes and confuses essential progress in our science. Importantly in our applied science, losing focus on sound scientific methods due to the misguided belief that it is required for regulatory expediency ultimately degrades our science and gradually lessens the usefulness of the associated evidence for solving environmental problems.

Additional issues can arise during the social processing of evidence. Chief among these challenges are biased weighting of opinions in dyads due to naïve realism, groupthink that occurs most often in homophilic groups, and polythink that emerges principally in heterophilic groups. Whether satisficing results in inadequate decision-making depends on context. For example, it is clearly undesirable within a scientific context but might be the only way forward to reach a regulatory decision.

A crucial process in any scientific field is the introduction of new evidence or concepts into the associated social group and subsequent movement of these innovations among individuals. Innovations move into a group at different rates, and these differences are only partially due to their relative scientific soundness. According to Kuhnian theory, the emergence and inevitable displacement of major scientific paradigms is natural, as is the associated resistance to radical paradigms by scientists vested in existing ones. Current ecotoxicology para-digms endangered by an increasing body of evidence include "Adverse ecological impacts of contaminants can be adequately predicted from collections of individual-based effect metrics" and "Strong, objective, and logically defensible inferences can be made with conventional null hypothesis significance test (NHST)–based procedures." Predictably, increasing levels of debate are taking place about the validity of these paradigms and possible replacement paradigms. In contrast to radical paradigms that precipitate Kuhnian crises, many incremental innovations cause minimal threat to vested parties and diffuse into ecotoxicology with little incivility from participants. The dynamics of most ecotoxicology innovations conform more to general innovation diffusion theory models than to Kuhnian theory.

The conventional internal or mixed-influence diffusion model has an initial period of slow acceptance until a certain tipping point is reached. The number of individuals engaged with the innovation then rapidly increases to a point at which it begins leveling off at a maximum number of successful adoptions. The 18 ecotoxicology innovations considered in this short book have not had enough time to move into the saturation phase but they exhibited lag and apparent linearly increasing phases. Since ecotoxicology emerged as a science, the lag time for innovation diffusion has become progressively shorter. Speculating, this shift suggests a transition from dynamics dominated by internal influences to one much more strongly controlled by external influences. Factors potentially contributing to this shift are the maturation of the science, emergence and increased impact of regulatory institutions, remarkable advances in IT and electronic access to information, and a steady increase in collaborative research. Regardless, an incremental innovation—good or bad—enters the field faster now than at any time in the past.

Successful introduction of an ecotoxicology innovation depends on several factors. In general, the initial step of acquisition from another group is less critical than the second step of effective integration of the innovation with existing methods and concepts in the field. Most innovation adoptions fail due to ineffective integration.

A conceptual or technical innovation moves in social networks and understanding network features can reveal patterns in innovation dynamics. As an example, the shortening of lag periods with time for the 18 ecotoxicology innovations was correlated with two network features. An increase in centrality (two-step reach) of innovators corresponded with a decrease in the length of time that an innovation spent in its lag phase. Also, innovation lag phase shortened as network cohesion increased with time. As expected, substantial homophily was characteristic of all explored co-authorship networks. However, the level of homophily was not correlated with the time until an innovation tipping point was reached. Regardless of the scientific soundness or regulatory utility of an innovation, fundamental network qualities are associated with adoption dynamics.

Most co-authorship networks for scientific innovations are characterized by a classic self-organizing core-periphery structure with a central group of highly productive individuals who co-author often together and a scattering of less prominent authors. Membership in the core can change through time, and less prominent members can engage with core members, potentially becoming core members themselves. This structure is superior to more static networks dominated by one or a few factions because members and information can move more effectively throughout the network.

9.3 THE USE OF ECOTOXICOLOGY EVIDENCE

It is of great use to a sailor to know the length of his line, though he cannot with it fathom all depths of the ocean.

Locke (1690)

Section 9.2 summarized the nature of ecotoxicological evidence, highlighting the common errors arising during evidence generation and evidence-based inferences. In terms of Locke's analogy as quoted above, the length of our line has been described at this point. What remains to be discussed briefly is enhancing the utility of evidence given these limits. As one important example, the reader should understand after reading Chapter 5, Individual Scientist: Reasoning by the Numbers, that a significant result from a conventional statistical test does not allow one to infer that an alternative hypothesis is true.[1]

Most importantly, ignorance of these limits weakens the usefulness of ecotoxicology evidence for critical activities, especially ecological risk assessment and remediation, and natural resource damage assessment and restoration. Recollect as examples the pervasive misunderstanding of how a *P*-value should be interpreted (see Chapter 5: Individual Scientist: Reasoning by the Numbers) and the gross misjudging by trained professionals of the likelihood that a pesticide is present in a tap water sample (see Chapter 2: Human Reasoning: Everyday Heuristics and Foibles). Those portrayed situations were not trivial or atypical. Substantial errors appeared that could be avoided by applying the concepts and techniques described in Chapters 2 and 5. The insights and suggestions made here for proceeding in the presence of

---

1. The integration of the Fisherian significance testing and Neyman–Pearson hypothesis testing vantages to produce the conventional NHST was discussed in general terms in Chapter 5, Individual Scientist: Reasoning by the Numbers (Section 5.2.2). Now that the concept of satisficing has been described, a more explicit description can be given. The conflict between groups in statistical sciences was eventually resolved via satisficing in which the Fisherian and Neyman–Pearson vantage were cobbled together into the current NHST paradigm. Illustrating the adverse consequences of satisficing during scientific processes, this led to the current incongruities described in Chapter 5, Individual Scientist: Reasoning by the Numbers, and the slow shift back toward sounder methods of statistical inference.

these kinds of foibles strengthen the usefulness and defensibility of evidence-based judgments.

In the first chapters, heuristics were described that can hinder sound decision-making from evidence. A solid understanding and consideration of these heuristics during ecological risk assessment will improve application of evidence to produce a defensible assessment. Construction of a checklist of heuristics and review of that list during deliberations might enhance conclusions. A review of impacts of heuristic use on existing ecotoxicology beliefs and risk assessment techniques might also expose flawed conclusions and allow their correction during deliberations.

Remnants of passé or flawed science practices such as precipitate explanation and weak testing occasionally appear in our science. They need to be recognized as no longer acceptable and dealt with before they corrupt the collective body of knowledge so painstakingly collected and organized by environmental scientists. The soundness of ecotoxicological evidence and inferences can be improved by appropriate adherence to the most promising rules described in Chapter 3, Human Reasoning: Within Scientific Traditions and Rules, as strong inference, multiple working hypotheses, and abductive inference, and by actively avoiding behaviors described in Chapter 4, Pathological Reasoning Within Sciences as pathological science.

Errors that emerge during the social processing of evidence can be reduced in several ways as indicated in Chapter 6, Social Processing of Evidence: Commonplace Dynamics and Foibles. Group decisions can be improved by thoughtful decision unit engineering, explicit empowering of members to be impartial and to voice concerns, assignment of a devil's advocate, and the incorporation of a cool-down period before a group reaches its final decision. Formal elicitation of judgments from experts can improve decisions even further but such elicitations must be designed carefully. Particularly important to successful elicitation are expert independence, appropriate calibration, and training.

Evidence and concepts enter and then move throughout groups such as ecotoxicologists as a function of factors other than scientific soundness. Understanding this can ultimately improve the effectiveness of their use. According to the strength of weak ties concept, an ecotoxicologists wishing to contribute new concepts or methods into their field might make a regular habit of reading outside their field and engaging people in other fields. As suggested in Chapters 2, 7, and 8, professionals in the late early to middle stages of their careers might take more responsibility for introducing radical innovations, perhaps acquired from another discipline during intentionally heterophilic activities. When introducing a useful innovation, much effort needs to be expended during the second step of innovation introduction, that being, effective integration of the innovation with existing concepts and approaches.

The introduction of more incremental innovations is needed in ecotoxicology, and contrary to common belief, will in most instances generate very modest risk to the innovator's professional status. A potential adopter presented with an increasingly popular innovation should spend time gauging why it appears so appealing. Is it popular at the moment because of its scientific soundness and potential for answering unresolved questions, or does it simply have several qualities that would favor adoption of any innovation? Similarly, a regulatory panel considering inclusion of an innovation as part of "current, best science" might ask the same question. Increased vigilance is especially important given the general shortening of the innovation diffusion lag phase described in Chapter 8, Evidence in Social Networks.

> *We to the place have come, where I have told thee*
> *Thou shalt behold the people dolorous*
> *Who have foregone the good of intellect.*
>
> Dante (1308)

In Dante's Canto III, Virgil brings Dante to the Gates of Hell, explaining the punishment of the souls outside the gates who were indifferent to good and bad thinking during their lives. Instead of being admitted to Hell, these apathetic souls were forced to race about in placard-carrying mobs, colliding with other mobs and reassembling helter-skelter under random placards while being stung by flying insects. It is my belief that if ecotoxicologists and risk assessors remain apathetic to the "off-topic" issues described in this book, they will find themselves addressing major 21st century environmental problems in a haphazard and unhelpful way. Illiberal slogans such as "Inference based on conventional NHST is good enough," "Ecosystems are too complex to understand so single-species effects metrics are good enough," "Regression, not hypothesis testing," or "Hypothesis testing, not regression" will remain scribbled on placards clutched by apathetic ecotoxicologists and risk assessors. Given the seriousness of these problems, I sincerely urge the reader to consider the points in this book during their professional activities.

REFERENCES

Locke, J. (1690). *An essay concerning human understanding (Vol1)*, compiled in 1959. In Alexander Campbell Fraser (Ed.), New York: Dover Publications.

Melville, H. (1851). *Moby Dick or the white whale*. London: Harper and Brothers Publishing.

Dante, A. (1308). *The inferno*. (Henry Wadsworth Longfellow Transl.). Reprinted on 2003 by Barnes & Noble Books, NY.

Appendix 1

Ecotoxicology Innovation Survey Methods

A.1.1 INNOVATION DATA SET

Chapters 7 and 8, "How Innovations Enter and Move Within Groups" and "Evidence in Social Networks," respectively, discuss some results from a heretofore unpublished literature analysis by M. C. Newman and Marcos Krull, a PhD graduate student at the College of William and Mary's Virginia Institute of Marine Science. (Because these data analyses have not been previously published, they should be cited as "co-authorship analyses by M. C. Newman and M. Krull as reported in Newman, M. C. (2018). *The Nature and Use of Ecotoxicology Evidence: Natural Science, Statistics, Psychology and Sociology.* Cambridge, MA: Academic Press/Elsevier.) In this study, co-authorship was analyzed for the following 18 ecotoxicological innovations. Some were recent innovations and others were more than 40 years old, and included conceptual and technical innovations. One innovation, the gill surface interaction model was also examined for comparative purposes (see Fig. 7.6) but was not included in the final comprehensive analyses in this study.[1]

1. **Additive Index** is an index for quantifying the joint action of toxicants in mixture. Marking and Dawson (1975) provide the following summary of the index: "The individual toxic contributions of poisons were

---

1. The gill surface interaction model (GSIM) was introduced by Pagenkopf (1983) and Pagenkopf, Russo, and Thurston (1974) to predict metal ion toxicity from metal influence on fish gill function. The premises of the GSIM were the following: acute metal toxicity to fish results from respiratory dysfunction of the gills, some dissolved metal species are more toxic than others, gill surfaces form complexes with metals and protons in waters, metal exchange between water and gill surface is fast, the gills have a finite ability to bind metals, and competitive inhibition occurs between hardness metal cations (Ca^{+2} and Mg^{+2}), protons, and toxic metals (Pagenkopf, 1983).

summed, and the additive toxicity was defined by a linear index for two chemicals in combination. This index expresses the toxicity quantitatively: zero indicates additive toxicity, negative values indicate less than additive toxicity, and positive values indicate greater than additive toxicity." Early authors include S. J. Broderius, V. K. Dawson, L. L. Marking, W. L. Mauck, and L. L. Smith. Informative publications are Broderius and Smith (1979), Dawson and Marking (1973), Marking & Dawson (1975), Marking (1977), and Marking and Mauck (1975).

2. The **Alkaline Unwinding Assay** estimates DNA damage in cells of exposed individuals based on the time-dependent partial unwinding of DNA under alkaline conditions. The ratio of double to total DNA after a certain time under alkaline conditions is estimated by fluorescence. The measured amount of double-stranded DNA is inversely proportional to the number of strand breaks present in DNA, that is, the amount of DNA damage. (Please note that the comet or single-cell electrophoresis assay is also based on DNA alkaline unwinding but was not included in this literature survey.) Early authors who introduced this technique into ecotoxicology include J. Bickham, C. Blaise, J. Everaarts, D. Nacci, L. Shugart, and C. Theodorakis. Informative publications on this topic are Davison (1966), Nacci and Jackim (1989), Rydberg (1975), and Shugart (1988a,b).

3. **Aquivalence** is an equilibrium criterion used in environmental models much as fugacity is used in models for volatile organic compounds (see **Fugacity Model** below). This criterion was thought to be needed for modeling contaminants with little or no vapor pressure such as metals, some organometals, inorganic ions, and some organic contaminants. Instead of vapor pressures, equilibrium criteria of activity were used in aquivalence models to model metal contaminant fate and distributions. Early authors include M. Diamond, D. Mackay, and P. M. Welbourn. Informative publications are Mackay (1989) and Diamond, Mackay, and Welbourn (1992).

4. **Biomagnification** is the increase in contaminant concentration from one trophic level (e.g., prey) to the next (e.g., predator) attributable to accumulation of contaminant from food. As a consequence, contaminant concentrations such as those of persistent organic pollutants (POPs) and methylmercury increase from the base to the apex of food webs. An authorship network analysis (1960−2013) identified approximately 2493 authors on this topic. Early authors who introduced this concept include D. W. Anderson, A. I. Bischoff, Rachel Carson, J. J. Hickey, E. G. Hunt, L. B. Hunt, P. A. Isaacson, G. M. Woodwell, and C. F. Wurster. Informative publications are Hickey and Anderson (1968), Hickey and Hunt (1960), Hickey et al. (1966), and Macek and Korn (1970), and Woodwell, Wurster, and Isaacson (1967).

5. **BLM (Biotic Ligand Model)** is a conceptual model that focuses on metal−ligand complexes including those formed at crucial sites on biological surfaces such as gill surfaces. It is used to imply or predict relationships between dissolved metal concentration or activity and bioavailability or effect to organisms. Early authors on this topic include H. E. Allen, D. M. Di Toro, D. G. Heijerick, C. R. Janssen, J. S. Meyer, and C. M. Wood. Informative publications are Di Toro et al. (2001), Erickson (2013), Meyer et al. (1999), and Santore, Di Toro, Paquin, Allen, and Meyer (2001).

6. **Dynamic Energy Budget—Toxicity (DEB$_{Tox}$)** is a theory-rich approach using energy budgeting for individuals as a central theme around which survival, growth, and reproduction under the influence of toxicants are modeled. Standard toxicity test data may be incorporated directly into the DEB$_{Tox}$ approach. Early authors on this topic include J. J. M. Bedaux and S. A. L. M. Kooijman. Informative publications are Bedaux and Kooijman (1994), Klepper and Bedaux (1997), Kooijman (1993), and Péry et al. (2002a,b).

7. **Diffusive Gradients in Thin Films (DGT) devices** are passive sampling devices that use metal diffusion coefficients through a hydrogel layer to estimate metal concentration and speciation in natural waters. It is composed of a membrane filter interfacing with the water media containing the metals, a diffusion layer hydrogel such as an acrylamide gel, and a receiving media such as an exchange resin to which the metals bind after diffusing though the hydrogel. The amount of metal bound in the receiving phase is notionally related to the labile concentration of metal in the water. Early authors on this topic include I. J. Allen, W. Davison, H. Zhang, and D. W. Grime. Informative publications are Allan et al. (2007), Davison, Grime, Morgan, and Clarke (1991), Davison and Zhang (1994), Harper, Davison, Zhang, and Tych (1998), Li, Zhao, Teasdale, John, and Wang (2005), and Zhang and Davison (1995).

8. **FETAX (Frog Embryo Teratogenesis Assay—*Xenopus*)** is a teratogenesis assay using embryos of the clawed frog, *Xenopus leavis*. Initially used to infer human teratogenic hazard, the assay was adopted to assess teratogenic hazard to nonhuman species. In the assay, eggs are exposed to different concentrations of contaminant for 96 hours. The proportions of exposed eggs dying and the proportions of living embryos with developmental abnormalities are scored for each concentration. A teratogenic index calculated with the resulting LC50 (mortality) and EC50 (malformation) is then used as a measure of teratogenic hazard. Early authors who helped to introduce this technique into ecotoxicology include J. A. Bantle, D. A. Dawson, J. N. Dumont, D. J. Fort, R. G. Epler, B. L. James, and T. W. Schultz. Informative publications are Dawson, McCormick, and Bantle (1985) and Greenhouse (1976).

9. **FIAM (Free-Ion Activity Model)** is a conceptual model that emphasizes the dominating role of the free metal ion activity in determining the bioactivity (e.g., uptake and toxicity) of all cationic metal species. Early authors who helped to introduce this concept into ecotoxicology include F. F. M Morel, G. I. Harrison, P. G. C. Campbell, and R. Roy. Informative publications are Andrew, Biesinger, and Glass (1977), Ardestani, Van Straalen, and Van Gestel (2015), Campbell (1995), Chakoumakos, Russo, and Thurston (1979), Pagenkopf (1983, 1974), Sunda and Guillard (1976), Sunda, Engel, and Thuotte (1978).

10. A **Fugacity Model** is a fugacity-based model used by ecotoxicologists to describe and predict the fate and distribution of volatile contaminants in the environment, including accumulation in biota. Fugacity, calculated from vapor phase concentrations in environmental components and expressed as a pressure, is the escaping tendency from one phase into another. It (f) is related to concentration (C) using a fugacity capacity constant (Z), $C = Zf$. Early authors include D. Mackay and S. Paterson. Informative publications are Mackay (1979), Mackay and Paterson (1981, 1982), and Mackay et al. (1983a,b, 1985).

11. **Global Distillation/Fractionation.** Global distillation is the process by which POPs such as volatile organochlorine compounds are distilled from warmer regions of use to cooler regions of the globe. Because POPs differ in their individual rates of degradation, vapor pressures, and lipophilities, global fractionation also occurs in which some POPs move more readily than others toward the polar regions. The net result is a redistribution of the different POPs from the Equator or site of origin toward the cold polar regions of the Earth. Early authors include T. F. Bidleman, K. Breivik, J. Dachs, E. Goldberg, T. Gouin, T. Harner, K. C. Jones, R. Lohmann, D. Mackay, S. D. Meijer, D. C. G. Muir, and F. Wania. Informative publications are Dachs et al. (2002), Goldberg (1975a), Ockenden, Sweetman, Prest, Steinnes, and Jones (1998), Li et al. (2010), Simonich and Hites (1995), Wania and Mackay (1996).

12. **Index of Biotic Integrity (IBI)** is a composite, multimetric index originally constructed by combining 12 qualities of fish communities of warm water, low-gradient streams to determine the extent of stream degradation. Incorporated information included species richness and composition, trophic characteristics, and abundance and fish condition. This index has been modified for diverse communities/taxonomic groups and aquatic systems as a general indicator of biotic integrity. However, this literature analysis focused solely on the fish-based index of biotic integrity. Early authors on this index include P. L. Angermeier, J. Breine, A. T. Herlihy, R. M. Hughes, J. Karr, J. D. Lyons, T. P. Simon, L. Tancioni, B. Vondracek, and T. R. Whittier. Informative publications on this index are Karr (1981) and Ruaro and Gubiani (2013).

13. **Microtox Assay/Test** is a simple and rapid bacterial assay in which a decrease in bacterial bioluminescence reflects toxic action. The reduction in light output is proportional to the toxicant concentration in the test solution. Early authors introducing this assay include A. A. Bulich, J. C. Chang, D. Dezart, B. J. Dutka, K. L. E. Kaiser, K. K. Kwan, F. R. Leach, N. Nyholm, J. Peterson, W. Slooff, and P. B. Taylor. Informative publications on this topic include Bulich (1979) and Johnson (2005).

14. **Mountain Cold-Trapping** is "the enrichment of certain persistent organic compounds in various media (e.g., soil, water, snow, or tissues) with increasing elevation up mountain slopes. The mechanism for cold trapping, as implied in its name, is compound transition out of the atmospheric gaseous phase and into solid phases such as particulates, water droplets (rain and fog), snow, and components of the land surface at the colder temperatures that prevail at higher elevations" (Newman, 2015). Informative publications are Blais et al. (2003), Carrera et al. (2002), Davidson et al. (2003), Grimalt et al. (2001a,b), Muir, Ford, Grift, Metner, and Lockhart (1990), and Wania and Westgate (2008).

15. **Mussel Watch.** In 1975, E. Goldberg proposed "a world mussel watch (utilizing *Mytilus edulis* and similar species) in which specimens from perhaps 100 coastal and open ocean sites would annually be analyzed for their concentrations of halogenated hydrocarbons, transuranics, heavy metals, and petroleum" (Goldberg, 1975b). Since then, the mussel watch concept has been adapted by various institutes and researchers to biomonitor spatial and temporal trends in contaminants. For the purpose of this survey, mussel watch programs are defined as those collecting bivalve tissue concentration data from coastal, and in some cases, large freshwater bodies. The intention of such monitoring is to detect or track contamination trends. One example is the US NOAA National Status and Trends Mussel Watch Program that began monitoring bivalves in 1986 from US coastal waters and the Canadian/US Great Lakes. Early authors who introduced this technique into ecotoxicology include J. W. Farrington, E. D. Goldberg, J. H. Martin, R. W. Risebrough, and J. L. Sericano. Informative publications are Goldberg (1975b) and Kimbrough, Johnson, Lauenstein, Christensen, and Apeti (2008).

16. The **PICT (Pollution-induced Community Tolerance)** concept describes the increase in tolerance to a toxicant(s) resulting from species composition shifts in the community, acclimation of individuals, and genetic changes in populations making up the community. Its presence is evidence of a community response to the toxic agent and can suggest which agent caused the response. It was developed initially, and is still studied, with microfloral communities such as periphyton and soil microorganisms. Early authors include H. Blanck, S. Molander, S.-A. Wangberg, W. Admiraal, J. Bloem, A. M. Breure, G. D. Greve, S.

Pesce, M. Rutgers, S. D. Siciliano, E. Smit, and P. van Beelen. Informative publications are Blanck, Wängberg, and Molander (1988), and Molander, Blanck, and Söderström (1990).

17. An **SSD (Species Sensitivity Distribution)**-like approach was first applied in the 1970s by the EPA to facilitate estimation of Water Quality Criteria. Later, the SSD approach was formulated for use in probabilistic ecological risk assessments and formalized mathematically by Kooijman (1987). The distribution of results from a set of single-species toxicity tests (i.e., LC50, EC50, or NOEC) is used to predict a Hazardous Concentration (HCp where p is a defined percentage of species). Early authors on this topic include T. Aldenberg, A. E. Boekhold, A. J. Hendriks, J. A. Hoekstra, S. A. L. M. Kooijman, K. M. Y. Leung, J. Notenboom, W. Slooff, M. G. D. Smit, R. Toy, M. T. P. Traas, and P. J. van den Brink. Informative publications are Cormier and Suter (2013), Mebane, Hennessy, and Dillon (2008), Stephen et al. (1985), Suter (2002), and van Straalen (2002).

18. **Trophic Dilution** is "the decrease in contaminant concentration as trophic level increases. Trophic dilution results from the net balance of ingestion rate, uptake from food, internal transformation, and elimination processes favoring the loss of contaminant that enters the organism as food" (Newman, 2015). Early authors include W. H. O. Ernst, Y. Hirao, C. C. Patterson, D. E. Reichle, R. I. Van Hook, N. M. van Straalen, and G. M. Woodwell. Informative publications are Gray (2002), Hirao and Patterson (1974), Kearns and Vetter (1982), Laskowski (1991), LeBlanc (1995), Litzke and Hübel (1993), Pokarzhevskii & van Straalen (1996), Reichle & Van Hook (1970), Reichle, Dunaway, and Nelson (1970), Swanson (1985), Tang and Simó (2003), van Straalen & Ernst (1991), Wan et al. (2005), and Woodwell (1968).

Publications about an innovation were gathered together from searches including the Web of Science, ProQuest, and Google Scholar data bases, and then compiled into a single RefWorks file. After eliminating duplicate entries and correcting errors in the original data bases, the results were transferred into a Microsoft Excel spreadsheet. Title, keywords, and abstract were examined to ensure that inclusion of each publication was appropriate. The resulting spreadsheet was then provided to a second person (M.K.) who independently rechecked and corrected it as needed. The final Excel spreadsheet was used to produce tables and matrices needed for the analyses described below.

A.1.2 INNOVATION DIFFUSION

Each spreadsheet was sorted by year, and the numbers computed of publications and (unique) authors added each year. Basic statistics and metrics were calculated from the results with the SAS statistical package (SAS Institute Inc., Cary, NC). Plots of the number of authors versus time for all innovations displayed a lag period followed by a change in slope (tipping point) after which the cumulative number of authors on the topic grew at a faster rate than during the lag phase. No innovation appeared to have had sufficient time since inception to plateau as per the conventional diffusion models, or to plateau and decline as per Kuhnian theory. As a result, the estimated duration of the lag period and also rate of increase in the number of authors' post-lag phase were extracted from the data sets. Marcos Krull developed R script using the R Package changepoint (https://github.com/rkillick/changepoint) to detect the point in time at which the slope changed in the time course of cumulative number of authors versus calendar year. A linear model was used to fit the number of authors or number of publications versus time after the tipping point.

A.1.3   CO-AUTHORSHIP NETWORKS

For this book, networks were analyzed only for the initial lag phase and also for the equivalent number of authors for the most recent years. Results for these periods were then analyzed as described below.

The following author attributes were tabulated by reviewing each publication.

1. Was the author the original innovator? (Y/N).
2. What country was indicated in the author's address?
3. From what region of the world was the author publishing? (Europe/ United Kingdom, North America, South/Central America, Africa, Asia, Australia/New Zealand/Tasmania, Middle East).
4. What was their professional affiliation? (Academic, consulting, government, industry, NGO)?

An attribute matrix file was generated for use with the UCINET V6.577 network analysis program (Borgatti et al., 2002). Valued adjacency matrices needed for the network analyses were generated from the Excel files using R script written by M. Krull. With these files and UCINET, whole and ego network qualities and graphics were produced.

REFERENCES

Allan, I. J., Knutsson, J., Guigues, N., Mills, G. A., Foullac, A.-M., & Greenwood, R. (2007). Evaluation of the Chemcatcher and DGT passive samplers for monitoring metals in highly fluctuating water concentrations. *Journal of Environmental Monitoring, 9,* 672−681.
Andrew, R. W., Biesinger, K. E., & Glass, G. E. (1977). Effects of inorganic complexing on the toxicity of copper to *Daphnia magna. Water Research, 11,* 309−315.

Ardestani, M. M., Van Straalen, N. M., & Van Gestel, C. A. M. (2015). Biotic ligand modeling approach: Synthesis of the effect of major cations on the toxicity of metals to soil and aquatic organisms. *Environmental Toxicology and Chemistry, 34*, 2194−2204.

Bedaux, J. J. M., & Kooijman, S. A. L. M. (1994). Statistical analysis of bioassays, based on hazard modeling. *Environmental and Ecological Statistics, 1*, 303−314.

Blais, J. M., Wilhelm, F., Kidd, K. A., Muir, D. C. G., Donald, D. B., & Schindler, D. W. (2003). Concentrations of organochlorine pesticides and polychlorinated biphenyls in amphipods (*Gammarus lacustris*) along an elevation gradient in mountain lakes of western Canada. *Environmental Toxicology and Chemistry, 22*, 2605−2613.

Blanck, H., Wängberg, S.-Å., & Molander, S. (1988). Pollution-induced community tolerance— A new ecotoxicological tool. In J. Cairns, Jr., & J. R. Pratt (Eds.), *Functional testing of aquatic biota for estimating hazards of chemicals, ASTM STP* (988, pp. 219−230). Philadelphia: American Society for Testing and Materials.

Borgatti, S. P., Everett, M. G., & Freeman, L. C. (2002). *Ucinet 6 for Windows: Software for social network analysis.* Harvard: Analytic Technologies.

Broderius, S. J., & Smith, L. L., Jr. (1979). Lethal and sublethal effects of binary mixtures of cyanide and hexavalent chromium, zinc, or ammonia to the fathead minnow (*Pimephales promelas*) and rainbow trout (*Salmo gairdneri*). *Journal of the Research Board of Canada, 36*, 164−172.

Bulich, A. A. (1979). Use of luminescent bacteria for dertermining toxicity in aquatic environments. In L. L. Marking, & R. A. Kimerle (Eds.), *Aquatic Toxicology, ASTM STP* (667, pp. 98−106). Philadelphia: American Society for Testing and Materials.

Campbell, P. G. C. (1995). Interactions between trace metals and aquatic organisms: A critique of the free-ion activity model. In A. Tessier, & D. R. Turner (Eds.), *Metal speciation and bioavailability in aquatic systems* (pp. 45−102). Chichester: John Wiley & Sons Ltd.

Carrera, G., Fernandez, P., Grimalt, J. O., Ventura, M., Camarero, L., Catalan, J., . . . Psenner, R. (2002). Atmospheric deposition of organochlorine compounds to remote mountain lakes of Europe. *Environmental Science & Technology, 36*, 2581−2588.

Chakoumakos, C., Russo, R. C., & Thurston, R. V. (1979). Toxicity of copper to cutthroat trout (*Salmo clarki) under different conditions of alkali*nity, pH, and hardness. *Environmental Science & Technology, 13*, 213−219.

Cormier, S. M., & Suter, G. W., Jr. (2013). A method for deriving water-quality benchmarks using field data. *Environmental Toxicology and Chemistry, 32*, 255−262.

Dachs, J., Lohmann, R., Ockenden, W. A., Méjanelle, L., Eisenreich, S. J., & Jones, K. C. (2002). Oceanic biogeochemical controls on global dynamics of persistent organic pollutants. *Environmental Science & Technology, 36*, 4229−4237.

Davidson, D. A., Wilkinson, A. C., Blais, J. M., Kimpe, L. E., McDonald, K. M., & Schindler, D. W. (2003). Orographic cold-trapping of persistent organic pollutants by vegetation in mountains of western Canada. *Environmental Science & Technology, 37*, 209−215.

Davison, P. F. (1966). The rate of strand separation in alkali-treated DNA. *Journal of Molecular Biology, 22*, 97−108.

Davison, W., Grime, G. W., Morgan, J. A. W., & Clarke, K. (1991). Distribution of dissolved iron in sediment pore waters at submillimeter resolution. *Nature, 352*, 323−325.

Davison, W., & Zhang, H. (1994). *In situ* speciation measurements of trace components in natural waters using thin-film gels. *Nature, 367*, 546−548.

Dawson, D. A., McCormick, C. A., & Bantle, J. A. (1985). Detection of teratogenic substances in acidic mine water samples using the frog embryo teratogenesis assay—*Xenopus* (FETAX). *Journal of applied Toxicology, 5*, 234−244.

Dawson, V. K., & Marking, L. L. (1973). *Toxicity of mixtures of quinaldine sulfate and MS-222 to fish. Investigations in Fish Control* (53, pp. 1−11). LaCrosse, WI: U.S. Bureau of Sports Fisheries and Wildlife.

Diamond, M. L., Mackay, D., & Welbourn, P. M. (1992). Models of multi-media partitioning of multi-species chemicals: The fugacity/aquivalence approach. *Chemosphere, 25*, 1907−1921.

Di Toro, D. M., Allen, H. E., Bergman, H. L., Meyer, J. S., Paquin, P. R., & Santore, R. C. (2001). Biotic ligand model of the acute toxicity of metals. 1. Technical basis. *Environmental Toxicology and Chemistry, 20*, 2383−2396.

Erickson, R. J. (2013). The biotic ligand model approach for addressing effects of exposure water chemistry on aquatic toxicity of metals: Genesis and challenges. *Environmental Toxicology and Chemistry, 32*, 1212−1214.

Goldberg, E. D. (1975a). Synthetic organohalides in the sea. *Proceedings of the Royal Academy of London B, 189*, 277−289.

Goldberg, E. D. (1975b). The mussel watch—A first step in global marine monitoring. *Marine Pollution Bulletin, 6*, 111.

Gray, J. S. (2002). Biomagnification in marine systems: The perspective of an ecologist. *Marine Pollution Bulletin, 45*, 46−52.

Greenhouse, G. (1976). The evaluation of toxic effects of chemicals in fresh water by using frog embryos and larvae. *Environmental Pollution, 11*, 303−315.

Grimalt, J. O., Fernandez, P., Berdie, L., Vilanova, R. M., Catalan, J., Psenner, R., ... Battarbee, R. W. (2001a). Selective trapping of organochlorine compounds in mountain lakes of temperate areas. *Environmental Science & Technology, 35*, 2690−2697.

Grimalt, J. O., Fernandez, P., & Vilanova, R. M. (2001b). Trapping of organochlorine compounds in high mountain lakes. *TheScientificWorld, 1*, 609−611.

Harper, M. P., Davison, W., Zhang, H., & Tych, W. (1998). Kinetics of metal exchange between solids and solutions in sediments and soils interpreted from DGT measured fluxes. *Geochimica et Cosmochimica Acta, 62*, 2757−2770.

Hickey, J. J., & Anderson, D. W. (1968). Chlorinated hydrocarbons and eggshell changes in raptorial and fish-eating birds. *Science, 162*, 271−273.

Hickey, J. J., & Hunt, L. B. (1960). Initial songbird mortality following a Dutch elm disease control program. *Journal of Wildlife Management, 24*, 259−265.

Hickey, J. J., Keith, J. A., & Coon, F. B. (1966). An exploration of pesticides in a Lake Michigan ecosystem. *Journal of Applied Ecology, 3* (Supplement), 141−154.

Hirao, Y., & Patterson, C. C. (1974). Lead aerosol pollution in the High Sierra overrides natural mechanisms which exclude lead from a food chain. *Science, 184*, 989−992.

Johnson, B. T. (2005). 1. Microtox® acute toxicity test. In C. Blaise, & J. F. Férard (Eds.), *Small-scale freshwater toxicity investigations* (Vol. 1, pp. 69−105). Houten: Springer Netherlands.

Karr, J. R. (1981). Assessment of biotic integrity using fish communities. *Fisheries, 6*, 21−27.

Kearns, P. K., & Vetter, R. J. (1982). Manganese-54 accumulation by *Chlorella* spp., *Daphnia magna* and yellow perch (*Perca flavescens*). *Hydrobiologia, 88*, 277−280.

Kimbrough, K. L., Johnson, W. E., Lauenstein, G. G., Christensen, J. D., & Apeti, D. A. (2008). *An assessment of two decades of contaminant monitoring in the nation's coastal zone. NOAA Technical Memorandum NOS NCCOS* (74, p. 105). Silver Spring, MD: NOAA/National Centers for Coastal Ocean Science.

Klepper, O., & Bedaux, J. J. M. (1997). Nonlinear parameter estimation for toxicological threshold models. *Ecological Modelling, 102*, 315−324.

Kooijman, S. A. L. M. (1987). A safety factor for LC50 values allowing for differences in sensitivity among species. *Water Research, 21*, 269–276.

Kooijman, S. A. L. M. (1993). *Dynamic energy budgets in biological systems. Theory and applications in ecotoxicology.* Cambridge: Cambridge University Press.

Laskowski, R. (1991). Are the top carnivores endangered by heavy metal biomagnification? *Oikos, 60*, 387–390.

LeBlanc, G. A. (1995). Trophic-level differences in the bioconcentration of chemicals: Implications in assessing environmental biomagnification. *Environmental Science & Technology, 29*, 154–160.

Li, W., Zhao, H., Teasdale, P. R., John, R., & Wang, F. (2005). Metal speciation measurement by diffusive gradients in thin films technique with different binding phases. *Analytica Chimica Acta, 533*, 193–202.

Li, Y.-F., Harner, T., Liu, L., Zhang, Z., Ren, N.-Q., Jia, H., ... Sverko, E. (2010). Polychlorinated biphenyls in global air and surface soil: Distributions, air-soil exchange, and fractionation effect. *Environmental Science & Technology, 44*, 2784–2790.

Litzke, J., & Hübel, K. (1993). Aquarium experiments with rainbow trout (*Oncorhynchus mykiss* Walbaum) and carp (*Cyprinus carpio* L.) to examine the accumulation of radionuclides. *Archiv für Hydrobiologie, 129*, 109–119.

Macek, K. L., & Korn, S. (1970). Significance of the food chain in DDT accumulation by fish. *Journal Fisheries Research Board of Canada, 27*, 1496–1498.

Mackay, D. (1979). Finding fugacity feasible. *Environmental Science & Technology, 13*, 1218–1223.

Mackay, D. (1989). Application of the QWASI (Quantitative Water Air Sediment Interaction) fugacity model to the dynamics of organic and inorganic chemicals in lakes. *Chemosphere, 18*, 1343–1365.

Mackay, D., & Paterson, S. (1981). Calculating fugacity. *Environmental Science & Technology, 15*, 1006–1014.

Mackay, D., & Paterson, S. (1982). Fugacity revisited. *Environmental Science & Technology, 16*, 654A–660A.

Mackay, D., Joy, M., & Paterson, S. (1983a). A quantitative water, air, sediment interaction (QWASI) fugacity model for describing the fate of chemicals in lakes. *Chemosphere, 12*, 981–997.

Mackay, D., Paterson, S., & Joy, M. (1983b). A quantitative water, air, sediment interaction (QWASI) fugacity model for describing the fate of chemicals in rivers. *Chemosphere, 12*, 1193–1208.

Mackay, D., Paterson, S., Cheung, B., & Neely, W. B. (1985). Evaluating the environmental behavior of chemicals with a level III fugacity model. *Chemosphere, 14*, 335–374.

Marking, L. L. (1977). Method for assessing additive toxicity of chemical mixtures. In F. L. Mayer, & J. L. Hamelink (Eds.), *Aquatic toxicity and hazard evaluation, ASTM STP* (634, pp. 99–108). West Conshohocken, PA: American Society for Testing and Materials.

Marking, L. L., & Dawson, V. K. (1975). Method for assessment of toxicity or efficacy of mixtures of chemicals. *U.S. Fish and Wildlife Service Investigations in Fish Control, 67*, 1–8.

Marking, L. L., & Mauck, W. L. (1975). Toxicity of paired mixtures of candidate forest insecticides to rainbow trout. *Bulletin of Environmental Contamination and Toxicology, 13*, 518–523.

Mebane, C. A., Hennessy, D. P., & Dillon, F. S. (2008). Developing acute-to-chronic toxicity ratios for lead, cadmium, and zinc using rainbow trout, a mayfly, and a midge. *Water, Air, & Soil Pollution, 188*, 41–66.

Meyer, J. S., Santore, R. C., Bobbit, J. P., Debrey, L. D., Boese, C. J., Paquin, P. R., ... Di Toro, D. M. (1999). Binding of nickel and copper to fish gills predicts toxicity when water hardness varies, but free-ion activity does not. *Environmental Science & Technology, 33*, 913−916.

Molander, S., Blanck, H., & Söderström, M. (1990). Toxicity assessment by pollution-induced community tolerance (PICT), and identification of metabolites in periphyton communities after exposure to 4,5,6-trichloroguauacol. *Aquatic Toxicology, 18*, 115−136.

Muir, D. C. G., Ford, C. A., Grift, N. P., Metner, D. A., & Lockhart, W. L. (1990). Geographic variation of chlorinated hydrocarbons in burbot (*Lota lota*) from remote lakes and rivers of Canada. *Archives of Environmental Contamination and Toxicology, 19*, 530−542.

Nacci, D., & Jackim, E. (1989). Using the DNA alkaline unwinding assay to detect DNA damage in laboratory and environmentally exposed cells and tissues. *Marine Environmental Research, 28*, 333−337.

Newman, M. C. (2015). *Fundamentals of ecotoxicology. The science of pollution* (4th Ed). Boca Raton: CRC/Taylor and Francis.

Ockenden, W. A., Sweetman, A. J., Prest, H. F., Steinnes, E., & Jones, K. C. (1998). Toward an understanding of the global atmospheric distribution of persistent organic pollutants: The use of semipermeable membrane devices as time-integrated passive samplers. *Environmental Science & Technology, 32*, 2795−2803.

Pagenkopf, G. K. (1983). Gill surface interaction model for trace-metal toxicity to fishes: Role of complexation, pH, and water hardness. *Environmental Science & Technology, 17*, 342−347.

Pagenkopf, G. K., Russo, R. C., & Thurston, R. V. (1974). Effect of complexation on toxicity of copper to fishes. *Journal of the Fisheries Research Board of Canada, 31*, 462−465.

Péry, A. R. R., Flammarion, P., Vollat, B., Bedaux, J. J. M., Kooijman, S. A. L. M., & Garric, J. (2002a). Using a biology-based model (DEBTOX) to analyze bioassays in ecotoxicology: Opportunities and recommendations. *Environmental Toxicology and Chemistry, 21*, 459−465.

Péry, A.R.R., Mons, R., Flammarion, P., Lagadic, L., & Garric, J. (2002b). A modeling approach to link food availability, growth, emergence, and reproduction for the midge, *Chironomus riparius*. *Environmental Toxicology and Chemistry, 21*, 2507−2513.

Pokarzhevskii, A. D., & van Straalen, N. M. (1996). A multi-element view on heavy metal bio-magnification. *Applied Soil Ecology, 3*, 95−98.

Reichle, D. E., Dunaway, P. B., & Nelson, D. J. (1970). Turnover and concentration of radionuclides in food chains. *Nuclear Safety, 11*, 43−55.

Reichle, D. E., & Van Hook, R. I. (1970). Radionuclide dynamics in insect food chains. *The Manitoba Entomologist, 4*, 22−32.

Ruaro, R., & Gubiani, É. A. (2013). A scientometric assessment of 30 years of the index of biotic integrity in aquatic ecosystems: Applications and main flaws. *Ecological Indicators, 29*, 105−110.

Rydberg, B. (1975). The rate of strand separation in alkali of DNA of irradiated mammalian cells. *Radiation Research, 61*, 274−287.

Santore, R. C., Di Toro, D. M., Paquin, P. R., Allen, H. E., & Meyer, J. S. (2001). Biotic ligand model of the acute toxicity of metals. 2. Application to acute copper toxicity in freshwater fish and *Daphnia*. *Environmental Toxicology and Chemistry, 20*, 2397−2402.

Shugart, L. R. (1988a). An alkaline unwinding assay for the detection of DNA damage in aquatic organisms. *Marine Environmental Research, 24*, 321−325.

Shugart, L. R. (1988b). Quantitation of chemically induced damage to DNA of aquatic organisms by alkaline unwinding assay. *Aquatic Toxicology, 13,* 43–52.

Simonich, S. L., & Hites, R. A. (1995). Global distribution of persistent organochlorine compounds. *Science, 269,* 1851–1854.

Stephen, C. E., Mount, D. I., Hansen, D. J., Gentile, J. R., Chapman, G. A., & Brungs, W. A. (1985). *Guidelines for deriving numerical national water quality criteria for the protection of aquatic organisms and their uses, PB8* (pp. 5–227049). Washington: US Environmental Protection Agency.

van Straalen, N. M. (2002). Theory of ecological risk assessment based on species sensitivity distributions. Suter, G. W. (2002). In L. Posthuma, G. W. Suter, & T. P. Traas (Eds.), *Species sensitivity distributions in ecotoxicology* (pp. 37–48). Boca Raton: Lewis Publishers.

Sunda, W. G., Engel, D. W., & Thuotte, R. M. (1978). Effect of chemical speciation on toxicity of cadmium to grass shrimp, *Palaemonetes pugio*: Importance of free cadmium ion. *Environmental Science and Technology, 12,* 409–413.

Sunda, W. G., & Guillard, R. R. L. (1976). The relationship between cupric ion activity and the toxicity of copper to phytoplankton. *Journal of Marine Research, 34,* 511–529.

Suter, G. W. (2002). North American history of species sensitivity distributions. In L. Posthuma, G. W. Suter, & T. P. Traas (Eds.), *Species sensitivity distributions in ecotoxicology* (pp. 11–17). Boca Raton: Lewis Publishers.

Swanson, S. M. (1985). Food-chain transfer of U-series radionuclides in a northern Saskachewan aquatic system. *Health Physics, 49,* 747–770.

Tang, K. M., & Simó, R. (2003). Trophic uptake and transfer of DMSP in simple planktonic food chains. *Aquatic Microbial Ecology, 31,* 193–202.

Van Straalen, N. M., & Ernst, W. H. O. (1991). Metal biomagnification may endanger species in critical pathways. *Oikos, 62,* 255–256.

Wan, Y., Hu, J., Yang, M., An, L., An, A., Jin, X., … Itoh, M. (2005). Characterization of trophic transfer for polychlorinated dibenzo-*p-dioxins, dibenzofurans, non-* and *mono-*ortho polychlorinated biphenyls in the marine food web of Bohai Bay, North China. *Environmental Science & Technology, 39,* 2417–2425.

Wania, F., & Mackay, D. (1996). Tracking the distribution of persistent organic pollutants. *Environmental Science & Technology, 30,* 390A–396A.

Wania, F., & Westgate, J. N. (2008). On the mechanism of mountain cold-trapping of organic chemicals. *Environmental Science & Technology, 42,* 9092–9098.

Woodwell, G. M. (1968). Radioecological concentration processes. *The Quarterly Review of Biology, 43,* 355–356.

Woodwell, G. M., Wurster, C. F., Jr., & Isaacson, P. A. (1967). DDT residues in an East Coast estuary: A case of biological concentration of a persistent insecticide. *Science, 156,* 821–824.

Zhang, H., & Davison, W. (1995). Performance characteristics of diffusion gradients in thin films for the *in situ* measurement of trace metals in aqueous solution. *Analytical Chemistry, 67,* 3391–3400.

Appendix 2

ResearchGate Ecotoxicologist Survey

A.2.1 GENERAL

Established in 2008, ResearchGate is a social networking site for scientists and researchers, which includes information on publication history and citations with numbers of ties (degree 1 and 2) with publication co-authors, followers on a scientist's ResearchGate page, and project collaborators and followers. A conventional h-index is calculated to reflect publication output and citation rate (Hirsch, 2007). As an example, an h-index of 30 would indicate that the individual scientist had 30 publications with 30 or more citations each. Several network scores are also included by ResearchGate. Notionally, an RG Score grossly gages an individual's scientific reputation as computed from how peers have received their work. The network score, Direct Reach is the total number of unique co-authors, collaborators, and ResearchGate publication and study followers one has on ResearchGate. The Indirect Reach is the sum of unique followers of one's co-authors and collaborators. (Chapter 8, Evidence in Social Networks, provides more background for understanding Direct and Indirect Reach which are simply summations of degree 1 and 2 linkages, respectively.) A final score, RG Reach, gages the potential visibility of one's work by summing the RG Direct and RG Indirect scores.

These metrics were extracted from the ResearchGate pages for 80 haphazardly chosen ecotoxicologists on February 10, 2017. Specific names were replaced by numbers to make the results anonymous.

A.2.2 TABULATION FOR 80 ECOTOXICOLOGISTS

The following data table includes (left to right) each ecotoxicologist's ID number, the year that they published their first paper, the number of years from the first publication to 2017, h-index, RG Score, RG Reach, Direct

Reach, and Indirect Reach. As explained in Chapter 8, Evidence in Social Networks, the rightmost column contains the regression residuals from a simple linear model of h-index predicted from the number of years that an ecotoxicologist was publishing in the field.

1	1986	31	27	41.93	7219	274	6945	− 1.6077
2	1981	36	35	38.39	5893	110	6003	2.8090
3	2008	9	14	37.86	6886	406	7292	1.1588
4	1985	32	32	38.82	3833	167	4000	2.6756
5	1977	40	31	40.26	8619	188	8431	− 4.0577
6	1997	20	11	24.39	2509	96	2413	− 9.7244
7	1989	28	22	31.79	4766	71	4695	− 4.4577
8	1991	26	32	38.57	10,683	217	10,466	6.9756
9	1970	47	90	52.94	47,123	1744	45,379	49.9257
10	1986	31	39	41.52	12,309	447	11,862	10.3923
11	1983	34	23	36.36	9772	217	9555	− 7.7577
12	1970	47	56	44.24	12,738	239	12,499	15.9257
13	1968	49	59	47.22	27,794	727	27,067	17.4924
14	1982	35	40	41.25	11,891	244	11,647	8.5256
15	1984	33	28	39.15	9865	207	9658	− 2.0410
16	1992	25	8	18.46	2187	18	2169	− 16.3077
17	1985	32	47	40.13	8617	318	8299	17.6756
18	1984	33	21	33.11	877	114	763	− 9.0410
19	1967	50	36	41.98	5830	197	5630	− 6.2243
20	1983	34	19	31.56	886	48	838	− 11.7577
21	1998	19	15	27.83	965	105	860	− 5.0078
22	1973	44	50	43.05	11,691	516	11,175	12.0757
23	1976	41	15	24.76	1392	46	1346	− 20.7743
24	1970	47	40	41.80	7375	281	7094	− 0.0743
25	1989	28	25	32.86	4509	154	4355	− 1.4577
26	1998	19	18	29.76	2120	177	1943	− 2.0078
27	1986	31	39	39.26	15,204	253	14,951	10.3923
28	1986	31	33	38.08	11,952	248	11,704	4.3923
29	1999	18	29	40.82	24,270	697	23,573	9.7089
30	2005	12	14	27.38	5215	88	5127	− 0.9911
31	1983	34	36	40.39	10,687	415	10,272	5.2423
32	1988	29	9	18.18	798	91	707	− 18.1744
33	1988	29	41	42.77	9867	293	9574	13.8256
34	1977	40	24	31.40	4109	52	4057	− 11.0577
35	2001	16	9	23.69	856	48	808	− 8.8578
36	1983	34	32	37.74	6173	164	6009	1.2423
37	1976	41	17	25.62	740	36	704	− 18.7743
38	1983	34	23	30.21	3388	191	3197	− 7.7577
39	1978	39	15	29.43	187	49	138	− 19.3410
40	1977	40	56	42.95	16,033	419	15,614	20.9423
41	1970	47	49	64.31	13,667	459	13,208	8.9257
42	1958	59	33	40.89	6096	143	5953	− 15.6743
43	1966	51	50	43.79	13,236	535	12,701	7.0590
44	1963	54	63	39.23	12,470	558	11,912	17.9090
45	1993	24	12	26.65	5163	114	504	− 11.5911
46	2010	7	11	28.14	2797	211	2586	− 0.4078
47	1976	41	10	24.35	1781	48	1733	− 25.7743

48	1970	47	35	39.79	10,751	280	10,471	− 5.0743
49	1994	23	25	34.00	15,768	151	15,617	2.1256
50	1977	40	48	44.01	72,971	1345	71,626	12.9423
51	1985	32	22	34.82	3334	213	3121	− 7.3244
52	1968	49	28	37.22	4145	57	4088	− 13.5076
53	1968	49	56	46.33	12,340	332	12,008	14.4924
54	1964	53	37	41.54	6341	182	6159	− 7.3743
55	1969	48	25	34.41	4638	68	4570	− 15.7910
56	1978	39	23	27.90	8959	149	8810	− 11.3410
57	1968	49	47	43.88	15,007	263	14,744	5.4924
58	1971	46	36	38.05	3623	180	3443	− 3.3577
59	1986	31	6	15.94	631	87	544	− 22.6077
60	1982	35	37	39.36	10,101	258	9843	5.5256
61	1980	37	51	45.14	15,488	581	14,907	18.0923
62	1979	38	17	30.93	5417	138	5279	− 16.6244
63	1980	37	41	39.82	29,904	523	29,381	8.0923
64	2011	6	7	26.48	2016	142	1874	− 3.6912
65	2011	6	4	9.37	297	19	278	− 6.6912
66	1986	31	33	38.08	11,952	248	11,704	4.3923
67	1983	34	30	36.10	8455	165	8290	− 0.7577
68	2009	8	9	28.14	3903	85	3818	− 3.1245
69	1986	31	19	28.46	868	16	852	− 9.6077
70	1990	27	59	45.34	13,813	636	13,177	33.2589
71	1985	32	50	47.68	14,622	291	14,331	20.6756
72	1985	32	68	47.79	19,543	762	18,781	38.6756
73	1958	59	33	46.26	2792	78	2714	− 15.6743
74	1962	55	25	33.93	2751	56	2695	− 20.8076
75	1986	31	29	37.37	10,431	307	10,124	0.3923
76	1982	35	48	44.86	19,687	572	19,115	16.5256
77	1973	44	20	30.68	10,158	100	10,058	− 17.9243
78	1982	35	27	37.59	11,775	330	11,445	− 4.4744
79	1985	32	25	34.10	2215	175	2040	− 4.3244
80	1988	29	32	35.42	16,571	291	16,280	4.8256

REFERENCE

Hirsch, J. E. (2007). Does the h index have predictive power? *Proceedings of the National Academy of Sciences of the USA*, *104*, 19193−19198.

Index

Printed in the United States
By Bookmasters